The Physics Lab Manual
Part 2

Experiments to Accompany
Physics 1502 and Physics 2611 Laboratories
at Youngstown State University

Department of Physics and Astronomy
Youngstown State University

The Physics Lab Manual Part 2

Experiments to Accompany Physics 1502 and Physics 2611 Laboratories at Youngstown State University
Youngstown State University | Department of Physics and Astronomy | Fall 2019

Printed in the United States of America
10 9 8 7 6
ISBN: 978-1-61740-797-0

Van-Griner Learning
Cincinnati, Ohio
www.van-griner.com

President: Dreis Van Landuyt
Project Manager: Janelle Lange
Customer Care Lead: Lauren Houseworth

Clymer 797-0 Su19
312605-323707
Copyright © 2020

Acknowledgments

The YSU Department of Physics and Astronomy acknowledges the following for their assistance with this manual: Sharon Shanks, Physics and Astronomy, who designed and produced this manuscript; Carl Leet, YSU Media Center, who took the majority of the photographs, designed several of the diagrams and provided production support; Jerry Fullum and the crew of the YSU Engineering Shop, who helped design and constructed the apparatus for Experiment 4; Michael Schueller, B.S., 1994, who assembled much of the apparatus; Richard Pirko, Physics and Astronomy, who provided a valuable suggestion for the timing device used in Experiment 3; and Douglas Fowler and Ted Seman, Physics and Astronomy, for their many useful suggestions for improvement.

—William Cochran
Summer 1995

Additional acknowledgments to Sheila Bishop and Doug Fowler for help with proofreading and suggestions for improvements during the second revision.

—Jan Clymer
Fall 2003

Acknowledgments to Kathy Durrell for writing the math review added in the third revision, the drawing of the vernier caliper in the introductory pages and for the revision of instructions for the force probe in Experiment 6.

—Jan Clymer
Fall 2005

Additional acknowledgments to Greg Diamantis for updated pictures throughout the manual.

—Jan Clymer
Fall 2012

Table of Contents

Notes

The Physics Laboratory

READ ME

This manual contains instructions for the performance and analysis of experiments designed to supplement Physics lecture courses at Youngstown State University. The instructions are intended to be self-explanatory; normally no additional instructions will be given to you by your laboratory instructor.

Begin the experiment as soon as all members of your group are present. Large apparatus will have been assembled for you at your laboratory table. A tray containing smaller apparatus can be obtained from your instructor; be certain to obtain the tray that matches your table number.

ATTENTION!

Your success in the laboratory will depend largely on your preparation before beginning the experiment.

The instructions for making some measurements are intentionally vague, and you are expected to devise appropriate methods. Your instructor will assist you by answering questions and making suggestions where appropriate, but is present to observe your progress and evaluate your performance as much as to provide assistance.

Some of the experiments will not require the full scheduled time to complete, and you are free to leave when you have finished your measurements. Some of the experiments, on the other hand, may be difficult to complete in the time allowed. In these cases work rapidly and do as much as you can. *Your success in the laboratory will depend largely on your preparation before beginning the experiment.* Read the instructions before coming to the laboratory, and read the relevant sections of your lecture textbook.

Before writing your first report, it is *essential* that you read and understand the section **Precision and Accuracy** in this manual.

The Laboratory Report

An important objective for the student in the introductory physics laboratory is to learn the conventional methods for reporting experimental results. A laboratory report need not be long, but must be coherent and concise. Each instructor will have individual preferences for various parts of the report, but in general most reports follow the same form. In the absence of modifications or additions from your instructor, the following guidelines will produce an acceptable report. The guidelines have been modeled on the form of research articles in professional journals, which means you may assume you are directing your report to readers who will have as much knowledge of the subject as you have; lengthy descriptions of laboratory apparatus and well-known theoretical derivations normally need not be included.

The report consists of a title page followed by four sections: the introduction, the data, the calculations, and the analysis, as follows:

TITLE PAGE

In addition to the title of the experiment, the title page should include (a) your name, (b) the names of your laboratory partners, (c) your laboratory table number (stenciled on the apron of the table), (d) the date the experiment was performed, and (e) the date the report was submitted for grading.

INTRODUCTION

The introduction to your report should be brief and concise, written as if you were explaining the experiment to someone who has the same level of knowledge as your own. Theoretical derivations of equations are necessary only if your instructor or the laboratory manual calls for them, or in the uncommon event that you have based the experiment on an idea of your own with which others will not be familiar. Similarly, you need not spend time drawing diagrams of the equipment or explaining how a standard piece of laboratory equipment operates unless you have introduced modifications of your own design. For example, in Section 1, Experiment 2, you may state simply *"The Behr Free-Fall Apparatus was used to record the positions of a freely-falling plummet at 0.02 second intervals"*; a drawing is unnecessary since the apparatus is illustrated twice in the laboratory manual.

Only three things need to be included in the introduction:

1. A statement of **what** was *measured*,

2. An explanation of **how** the measurements were taken (and to what precision), and

3. A statement of the **purpose** of the measurements.

For example, suppose you were given a ruler and a block of wood approximately 100 cm long, 10 cm wide, and 1 cm thick and told to determine the volume of the block. An acceptable introduction would read, in part, something like this:

> *The length [**what**] of a block of wood was measured by means of a ruler [**how**] to a precision of one part in two thousand [see discussion below], the width to one part in two hundred, and the thickness to one part in twenty in order to determine the volume [**purpose**] of the block by means of the equation Volume = Length × Width × Thickness.*

No diagram needs be given of either the ruler or the block, and no derivation of the equation for the volume is required. Note that it is the length, width, and thickness that are *measured*, and the volume is *determined* from the measurements. The term precision as used in step (b) refers to the precision a given measuring device is capable of providing *relative to the size of the measurement being taken*. In the example it has been assumed that the ruler was divided into millimeter divisions and that estimates were made to the nearest half millimeter. Suppose then that this ruler is used to measure the length of the block, approximately 100 cm or 1000 mm, to the nearest 0.5 mm; 0.5 mm is 5/10000 or 1/2000 of 100 cm, so the precision is expressed as one part in 2000. Similarly, 0.5 mm is 5/1000 or 1/200 of 10 cm, so the relative precision of the width measurement is expressed as one part in 200, and for the thickness 0.5 mm is 5/100 or 1/20 of one cm and the relative precision is expressed as one part in 20. This method of specifying the precision reflects the fact that measuring a long length with such a ruler is more precise than measuring a short length; the ruler would provide a relatively precise determination of the length of a 10-meter laboratory bench, but would not be a very good device to use to measure the thickness of a fingernail.

ATTENTION!

Laboratory work is, by nature, cooperative; data are taken as a group effort. But it should be clear that submission of work done by another, use of data acquired by another without permission and acknowledgment, and fabrication and alteration of data to fit a "desired" result are dishonest acts and will result in a course grade of F and the submission of a report of academic dishonesty to the student's dean and to the Associate Vice President for Student Services.

DATA

Data should be recorded in ink on standard notebook paper during the course of the experiment. The data should be recorded in a coherent form as measurements are taken so that recopying for the laboratory report is unnecessary. Initial and date the data sheet. *The data sheet should be reviewed **and signed by your instructor** before you leave the laboratory.*

CALCULATIONS

Calculations should be presented in *tabulated* form; there are examples in the **Precision and Accuracy** section of this manual as well as in Section 1, Experiment 2.

ANALYSIS

The analysis section is the most important section of the laboratory report. It need not be lengthy, but should present a *quantitative* discussion of the results. The emphasis should be on the sources of the imprecision reported, i.e., on the quantitative extent to which your measurements were able to illustrate or verify the particular physical law being studied. Any questions asked in the laboratory instructions should also be answered in this section of the report. *Specific* sources of imprecision or error should be discussed *quantitatively*, and vague generalizations are to be avoided. Further discussion of these points is given in the **Precision and Accuracy** section of this manual.

Notes

Precision and Accuracy in Measurement: A Guideline

INTRODUCTION

Approximation is inherent in measurement, and no measured parameter can ever be known exactly. For this reason an experimental result is considered complete only when a quantitative expression is given of the precision of the result, a statement is made that delineates the sources of the imprecision, and an analysis is presented of the relative contribution of each source. These notes are intended as an abbreviated guide to the conventional methods used to report experimental results. It is expected that you will refer to the assigned text on data analysis for a more thorough treatment.

TERMINOLOGY

Terminology used in error analysis is sometimes ambiguous. Consider for example these definitions from the *Random House Dictionary of the English Language:*

> ER/ROR: The difference between the observed or approximately determined value and true value of a quantity.

> UN/CER/TAIN: Not clearly or precisely determined; indefinite.

According to the above definition of error, the term applies only to a comparison between the true value of a quantity and the experimentally determined value of the quantity. But usually there is no "true" or "known" value for a quantity being measured; determination of the value is the purpose of the measurement. In the physical sciences the term error, in fact, is used to mean two different things: error in the dictionary sense, and as an expression of the uncertainty, or imprecision, in an experimental result. The meaning of the term must be judged from context.

The sketches in Figure 1 illustrate the difference between the concepts of accuracy and precision. The four sketches represent marksmanship targets; the "true" or "correct" result for the marksman would be striking the bullseye, but this does not happen on every firing of the weapon. In Figure 1 (a) the marksman has obtained a precise result, for all firings have produced a closely grouped cluster of marks (i.e., the uncertainty is small), but the accuracy is poor (i.e., the error is high) since the cluster is not centered near the bullseye. Figure 1 (b) represents both low precision (large uncertainty) and poor accuracy (high error), and Figure 1 (c) high precision and high accuracy. What about Figure 1 (d)?

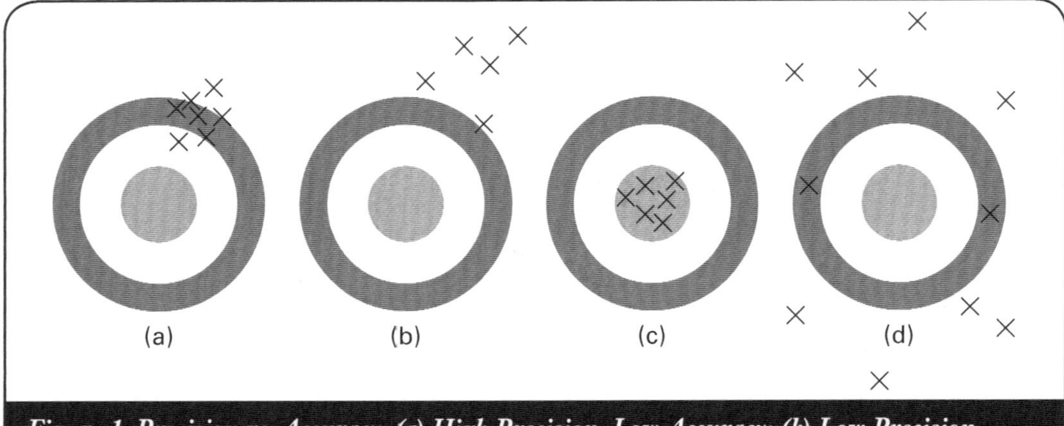

Figure 1: Precision vs. Accuracy. (a) High Precision, Low Accuracy; (b) Low Precision, Low Accuracy; (c) High Precision, High Accuracy; (d) ?

AN EXAMPLE

In the early part of this century physicists were studying the character of light emitted by a heated object. Metal, for example, glows red when heated; heated further the color changes, and when a sufficiently high temperature is reached the metal appears nearly white. Such emissions were studied carefully, the relative amounts of energy emitted at different frequencies (colors) being the measured quantity. A careless investigator might have graphically presented the result as shown in Figure 2.

At the time the measurements were being made, a theory had been developed to predict what the results should look like. Presented graphically, this theoretical prediction would look something as shown in Figure 3.

A comparison of Figures 2 and 3 will show an apparent agreement at smaller frequencies but an apparent disagreement at higher frequencies. But, in fact, it's not clear that there is any correlation between these two graphs because the investigator did not indicate how uncertain the experimental results were. Suppose instead the experimental graph had shown the uncertainty in the results by including with each point a small "error" bar. If the results had looked like Figure 4 (the theoretical plot has been superimposed on the data to make comparison easier), then there could be no doubt that a serious discrepancy exists between theory and experiment. This is in fact what happened. As a

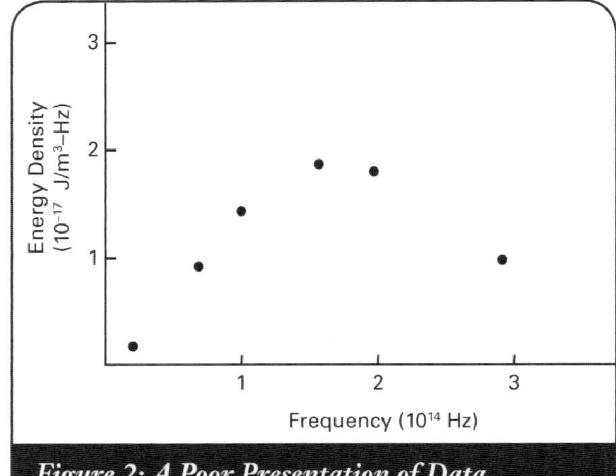

Figure 2: A Poor Presentation of Data

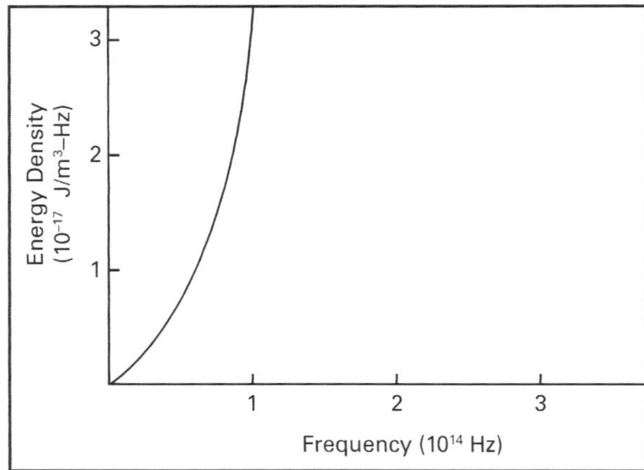

Figure 3: A Theorized Prediction of Results in Figure 2

result, a new look was taken at the theory, and the required modifications formed the basis of the quantum theory of radiation.

On the other hand, if the experimental results had looked like Figure 5, then the theoretical curve could still conceivably be consistent with the experimental results. What would then become crucial would be a knowledge of what physically caused the experimental uncertainties to be so large. The experiment could then be redesigned to reduce these uncertainties, and a more meaningful comparison of theory and experiment could be made.

THE MATHEMATICS
DISTRIBUTION IN RANDOM ERROR

When a parameter is to be measured the measurement is repeated many times in order to achieve an average value, the principle being that random fluctuations in the result are as likely to produce a value somewhat larger than the average as they are to produce a result somewhat smaller.[1] There are in general many possible causes of the fluctuations, such as estimation of scale readings by the observer, irregularities in the parameter being measured, etc. The fluctuations obtained can be illustrated by means of a *histogram*. For example, in Experiment 1 the time for the sand to empty from a sandglass timer is measured using an electric stop clock capable of recording time intervals to the nearest hundredth of a second. Repetitive measurements might produce results something like this:

Trial #1:	181.25 seconds
2:	180.05
3:	178.95
4:	179.55
5:	180.05
6:	180.05
.	.
.	.
100:	178.95
.	.
Average:	180.00

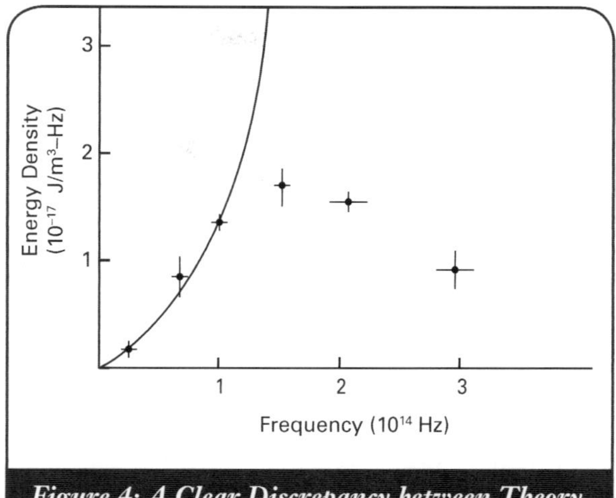

Figure 4: *A Clear Discrepancy between Theory and Experiment*

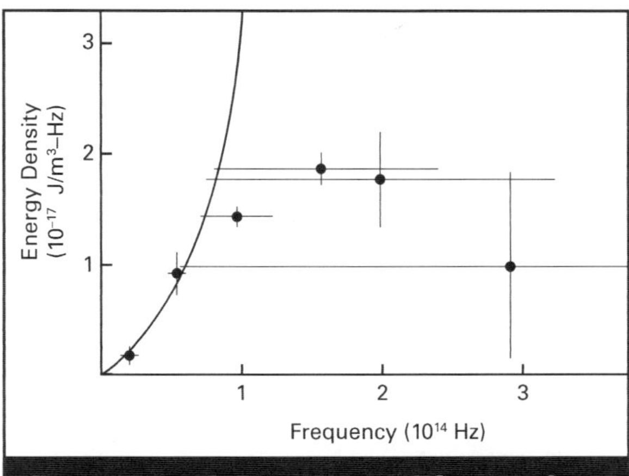

Figure 5: *Result Requiring Modification of Experimental Procedure*

Note

[1] See Section 1.4 in Lyons, Louis, *A Practical Guide to Data Analysis for Physical Science Students*, Cambridge: Cambridge University Press, 1991.

The histogram is constructed by drawing a vertical and a horizontal axis and dividing the horizontal axis into time second, centered about the average of the measured values. For each interval on the horizontal axis, the number of times a measured value was obtained which lay in that particular interval is plotted on the vertical axis. The result will look something like Figure 6.

If the distribution of values about the average is due to random causes, then if the size of the intervals on the horizontal axis is made smaller and smaller, the histogram will acquire a symmetric shape, and the points defining the centers of the tops of each vertical bar in the histogram will define a smooth curve, the *normal* or *Gaussian* distribution. This can be shown mathematically, and is known as the Central Limit Theorem. Such a curve is shown in Figure 7. The vertical axis must be reinterpreted as a representation of the *probability* that the corresponding value on the horizontal axis will be obtained on any given measurement, and the scale of the vertical axis is chosen so that the area under the curve has unit value.

The average value of the measured quantity is the numerical value on the horizontal axis directly under the peak of the distribution

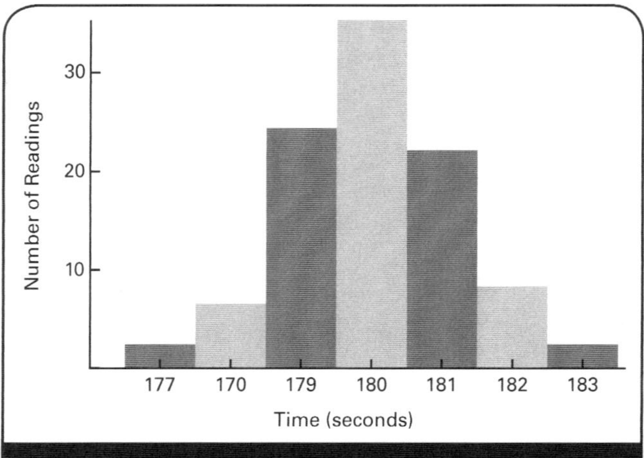

Figure 6: Histogram Showing History of Occurrence of Time Values out of a Total of 100 Readings

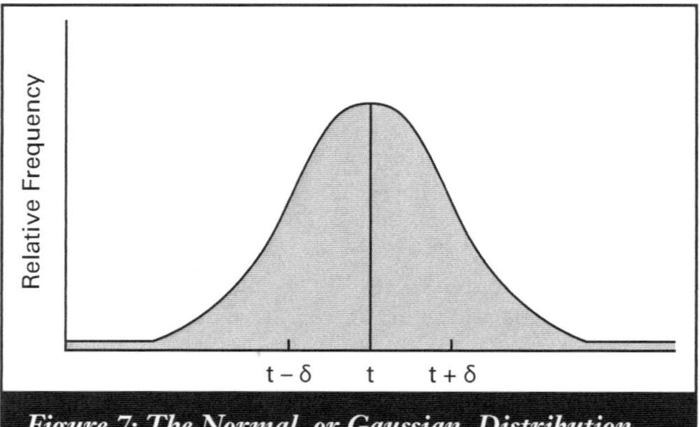

Figure 7: The Normal, or Gaussian, Distribution

curve, and clearly the width of the curve is related to how precisely the quantity has been determined. Convention dictates that the width of the curve is specified by the distance between the two points on either side of the average value which fall where the curvature of the distribution changes from concave to convex (some convention must be established to specify the width, since the curve never falls to the horizontal axis). It turns out that the width between the points where the curvature changes can be calculated from the table of experimental results. To the table is added a second column representing the difference (or deviation) between each experimental value and the average:

Trial #1:	$t = 181.25$ seconds	deviation $= 1.25$ seconds
2:	180.05	0.05
3:	178.95	-1.05
4:	179.55	-0.45
5:	180.05	0.05
6:	180.05	0.05
.	.	.
.	.	.
.	.	.
100:	178.95	-1.05
.	.	.

The standard (or rms) deviation, which is the distance on the horizontal axis of the Gaussian distribution curve from the average value to the change of curvature point on either side of the average (Figure 7) is found by squaring each value in the deviation column, adding the squares, dividing by the number of entries in the table,[2] and taking the square root of the result, i.e.,

$$\delta_t = \sqrt{\frac{\Sigma_i (t - \overline{\overline{t}})^2}{n}}$$

where t_i is the ith experimental value, $\overline{\overline{t}}$ the average value, and n the number of measurements taken. The experimental result is then reported as

$$t = \overline{\overline{t}} + \delta_t$$

In fact in many experimental measurements the above calculation is *not* justified: For the histogram of Figure 6 to transform into the Gaussian distribution of Figure 7 *many* measurements would have to be represented. How many? The answer is subjective, and the following guidelines may be used to determine how best to present a quantitative measure of the precision of an experimental result:

1. If more than 50 trials have been taken, represent the uncertainty by the rms deviation.

2. If more than 10 but fewer than 50 trials have been taken, report the uncertainty as the *average of the absolute values* of the quantities in the deviation column. (Why absolute values?)

3. If fewer than 10 measurements have been taken, report the uncertainty as the largest of the individual deviations.

4. If only one or two experimental values are recorded, the observer must make a reasonable estimate of the uncertainty, based on experience with the measuring apparatus and the characteristics of the object being measured.

PROPAGATION OF UNCERTAINTY

Often in experimental work the quantity that is measured is not the result sought, but rather is used along with other measured quantities to *calculate* the desired result. Since each measured quantity is uncertain by some amount there will be uncertainty in the calculated result as well. Propagation of uncertainty deals with how the uncertainty in the calculated result is determined from the uncertainties in the measured quantities. The correct procedures follow from calculus, but the following approximate methods will suffice for most calculations in this laboratory:

If a quantity x is to be calculated from two experimentally determined parameters y and z, with average results $\overline{y}$ and $\overline{z}$ and respective uncertainties δ_y and δ_z, then

    a. for addition or subtraction,
$$x = \overline{y} + \overline{z} \quad or \quad x = \overline{y} - \overline{z}$$
$$\delta_x = \delta_y + \delta_z$$

    (i.e., the uncertainty in *x* is the sum of the uncertainties in *y* and *z*)

Note
[2] See Section 1.4 in Lyons, Louis, *A Practical Guide to Data Analysis for Physical Science Students*, Cambridge: Cambridge University Press, 1991.

b. for multiplication or division,

$$\bar{x} = (\bar{y})\,(\bar{z}) \quad or \quad x = \frac{\bar{y}}{\bar{z}},$$

$$\frac{\delta_x}{x} = \frac{\delta_y}{y} + \frac{\delta_z}{z},$$

(i.e., the fractional uncertainty in x is the sum of the fractional uncertainties in y and z);*

c. for powers and roots,

$$x = \bar{y}^n,$$

$$\frac{\delta_x}{x} = n\left(\frac{\delta_y}{y}\right)$$

d. for logarithmic relations,

$$x = \log \bar{y},$$

$$\delta_x = (.434)\frac{\delta_y}{\bar{y}}$$

NOTE!

All uncertainties and fractional uncertainties should be taken as positive. The percent uncertainty for any quantity can be found by multiplying the fractional uncertainty by 100.

e. for trig functions,
If $x = \sin y$, $\delta x = \cos y\ \delta y$.
If $x = \cos y$, $\delta x = \sin y\ \delta y$.
If $x = \tan y$, $\delta x = \sec^2 y\ \delta y = \delta y\,/\cos^2 y$.

The five rules given above are approximations to results derived from calculus and are sufficient for the calculations required for this laboratory. A more complete discussion of error propagation can be found in *A Practical Guide to Data Analysis for Physical Science Students*, Cambridge University Press, 1991.

Note
The factor of .434 arises from the fact that $\log y = (\log e)(\ln y)$, so $\delta(\log y) = (\log e)\ \delta(\ln y)$. Since $\log e$ $\approx .434$, this gives $\delta(\log y) \approx (.434)\frac{\delta_y}{y}$

SIGNIFICANT FIGURES

When doing simple calculations with electronic calculators or computers there is often temptation to report all the digits registered by the calculator, but it should be clear that the reporting of more figures than are justified by the uncertainty in the specified quantity is meaningless. For example, if in the earlier example of the sandglass measurement the value obtained by averaging over all readings is 180.78 seconds, yet the resultant rms deviation turns out to be 2 seconds, it makes no sense to report the experimental result as 180.78 ± 2 seconds. If it is the first figure to the left of the decimal point that is uncertain, then there can be no significance to the two figures to the right of the decimal point. The correct expression of the result would be 181 ± 2 seconds.

When combining numbers to calculate a result, two guidelines should be followed:

    a. When adding or subtracting, keep no more decimals in the result than are in the parameter having the fewest decimals. For example,

$$26.135 + 0.31 = 26.45$$

    b. When multiplying or dividing, the result will have the same number of significant figures as the factor that has the smallest number of significant figures. For example,

$$(315.26) \times (66.8) = 211 \times 10^4$$

PERCENT ERROR

In rare instances there is a generally accepted value for the quantity that is to be determined experimentally. In such cases the error in the experimental result is reported as a percent error:

$$\% \ error = \frac{|accepted \ value - experimental \ value|}{accepted \ value} \times 100$$

Then the analyst must ask whether the accepted value, if not in agreement with the mean experimental value, lies within the reported range of uncertainty of the experimental value and if not, why not. An important point to keep in mind is that the result reported by the laboratory observer *is* the correct result for the experiment as performed under the given laboratory conditions. Errors are not made by carelessness or forgetting to take an important measurement; citing carelessness as a source of error is not acceptable. The source of error should rather be sought in a difference between the experimental conditions under which the observer worked and the conditions which existed when the "accepted" value was determined. Such conjectures should be considered *quantitatively*. For example, a piece of wire will have different lengths when measured at different temperatures; knowing the coefficient of thermal expansion of the wire, the temperature difference required to produce the reported error should be calculated and compared with the known temperature difference.

COMPARISONS

As stated above, only rarely is the laboratory observer provided with an accepted value with which the experimental result can be compared. Often, however, the same parameter will be determined by two or more different methods in an effort to improve precision and detect possible *systematic* errors. Then the different results are compared using a percent difference:

$$\% \ difference = \frac{|experimental \ value \ 1 - experimental \ value \ 2|}{average \ of \ experimental \ values \ 1 \ and \ 2} \times 100$$

In such a case an analysis of the relative precisions of the methods should be made; physically why did one method yield a result with more (or less) uncertainty than the other? Such considerations must be quantitative.

GOOD VS. POOR RESULTS

Can a result be characterized as good, poor, fair, etc.? The answer is that it is the use to which the results will be put that determines whether the measurement has been sufficiently precise. Knowing the speed of a runner to the nearest foot per second may be sufficient for predicting the time for a jog around the block, but insufficient for determining the result of a track meet. The only good analysis of an experiment is a quantitative analysis, and subjective generalizations should be avoided. Let the numbers speak for themselves.

Using a Vernier Caliper

A **vernier caliper** is a device that allows a precise measurement around an object (using the jaws) or of the space between two edges (using the arms). (A vernier can also be used for depth measurements, but we won't make use of that feature in this course.) See Figure 8.

To measure the length of an object, loosen the thumbscrew on the top of the vernier and insert the object between the jaws of the vernier. Then move the slide roller to adjust the gap between the jaws until it fits precisely the dimension you wish to measure. You will notice that there are two separate scales along the bottom of the vernier. (There are two scales along the top also, but these scales use British units, so ignore them.) There is a fixed scale that is marked in centimeters, with smaller markings to indicate tenths of a centimeter (i.e., mm). In addition, there is a sliding scale with eleven markings. This is the vernier scale. Take a reading from the fixed scale at the point where the leftmost mark of the vernier scale abuts it. If the mark on the vernier is between two marks on the fixed scale above it, use the smaller of the two. So, for instance, in Figure 9 on the next page, the leftmost mark on the vernier scale is between 1.1 and 1.2 on the fixed scale, so we use 1.1. (See A in the diagram.) Next examine the caliper carefully and determine which of the ten markings beyond the leftmost one on the vernier scale lines up exactly with a marking on the fixed scale above. This number gives the reading for the hundredths place for your measurement. In Figure 9, the sixth line on the vernier scale lines up exactly with the line on the fixed scale above it. (See B in the diagram.) So the value you would record for this measurement would be 1.16 cm.

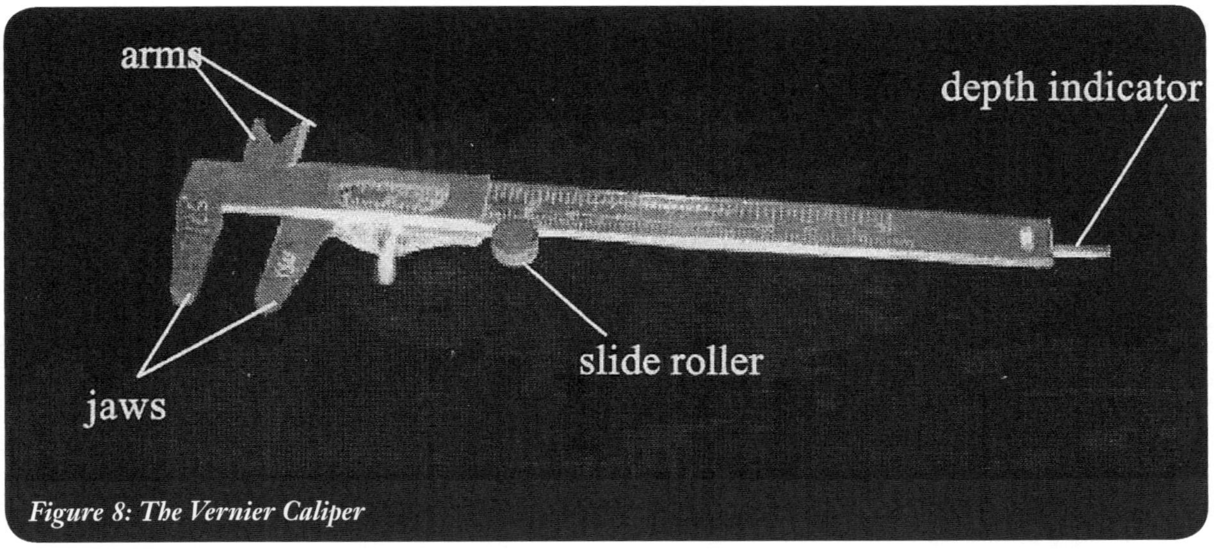

Figure 8: The Vernier Caliper

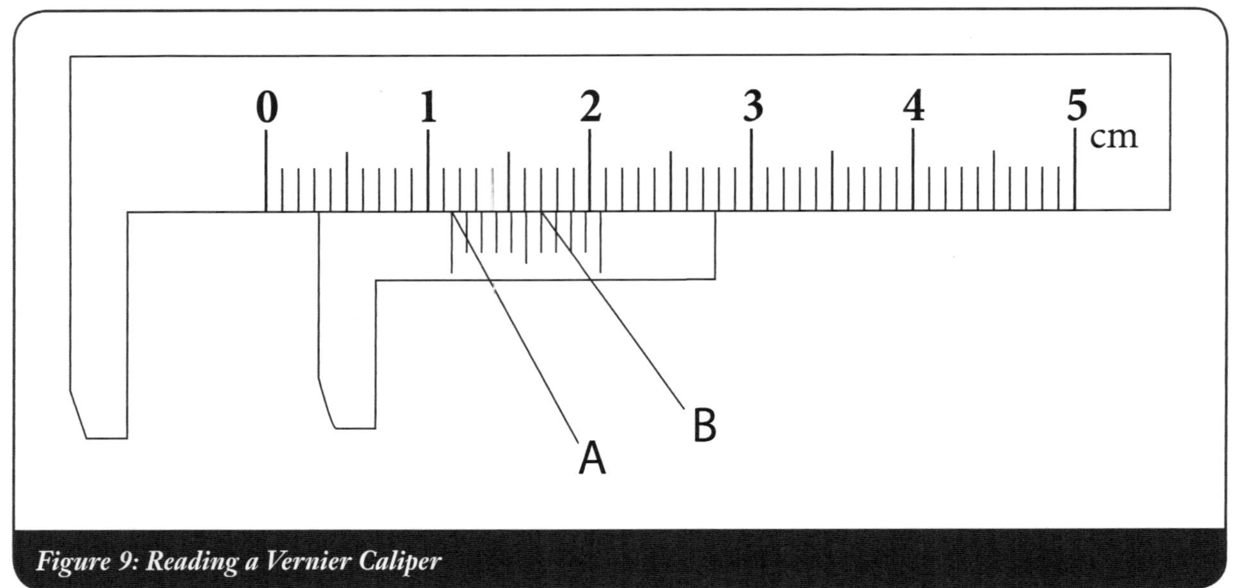

Figure 9: Reading a Vernier Caliper

For practice using a vernier caliper, go to *http://www.wcsscience.com/vernier/caliper.html*. (You will notice that the vernier scale there is labeled in mm. With a little thought, it should be evident that there is no difference between that and the cm label on the main scale on the photograph and the diagram above.)

Math Review

There are some basic math skills you will need in order to succeed in this lab. This section is intended to be a review of those skills. If any of the topics presented here are unfamiliar to you, you should consult a math text to get a more complete explanation of that topic.

BASIC ALGEBRA AND MANIPULATION OF EQUATIONS

If
$$a = \left(\frac{b}{c}\right)$$

then
$$b = ac \quad \text{and} \quad c = \frac{b}{a}$$

If we also know that $a = bd$, then by combining the two equations:

$$a = bd = \frac{b}{c}$$

and
$$d = \frac{1}{c} = c^{-1}$$

ADDING THE INVERSE OF TWO NUMBERS

$$a^{-1} + b^{-1} = \left(\frac{1}{a}\right) + \left(\frac{1}{b}\right)$$

$$= \left(\frac{b}{b}\right)\left(\frac{1}{a}\right) + \left(\frac{a}{a}\right)\left(\frac{1}{b}\right)$$

$$= \frac{(a+b)}{ab}$$

Note that this is **NOT** the same as $\dfrac{1}{(a+b)}$

POWERS

If a number (a) raised to a power is multiplied by the SAME number raised to another power, the exponents add. For example,

$$a^2 a^3 = a^{(2+3)} = a^5$$

Similarly if a number raised to a power is divided by the SAME number to another power the exponents are subtracted:

$$\frac{a^5}{a^2} = a^{(5-2)} = a^3$$

This is NOT THE CASE if the bases are different.
i.e.,

$$a^2 b^3 \neq (ab)^5$$

LOGARITHMS

The two most commonly used logarithms are the base 10 logarithm (usually expressed as log or $\log_{10}$) and the natural logarithm (expressed ln) which is base e, where e = 2.7182818…. Both the common log and the natural log behave in exactly the same way in any manipulations. The only difference between the two is the base. You can also create a log in any base you choose; however, only base 10 and base e logs will be used in this lab.

The common log is defined by

$$\log a = n \iff 10^n = a$$

i.e., the common log of a is the power to which 10 must be raised to obtain a. (The notation "log" is understood to mean $\log_{10}$, i.e., base 10 log.)

Similarly the natural (base e) log is defined by

$$\ln a = n \iff e^n = a$$

meaning that the natural log of a is the power to which e (2.7182818…) must be raised to obtain a. (The "ln" notation is understood to mean $\log_e$, i.e., base e log.)

Taking a log is the inverse of raising the base to a power. So, $10^{\log(n)} = n$ and, conversely, $\log(10^n) = n$.

The log of a PRODUCT of two numbers is equal to the SUM of the logs of the two numbers i.e.,

$$\log (ab) = \log (a) + \log (b)$$

Similarly the log of the QUOTIENT of two numbers is the DIFFERENCE of their logs i.e.,

$$\log \left(\frac{a}{b} \right) = \log (a) - \log (b)$$

Now since a^n is just a multiplied by itself n times,

$$\log (a^n) = n \log (a)$$

If we have a power-law equation, $y = a x^n$, we can express this in the form of the equation for a straight line by taking the log of both sides:

$$\log y = \log (ax^n)$$

$$\log y = \log a + \log x^n$$

$$\log y = \log a + n \log x$$

$$\text{or} \quad \log y = n \log x + \log a$$

which has the same form as the equation of a straight line: $y = m x + b$ if we replace y by log(y) and x by log(x) as shown below. The slope of this line is then given by n and the y-intercept is given by log (a).

$$\log y = n \log x + \log a$$

$$y = m \quad x \quad + \quad b$$

This is a general result. Whenever y can be expressed as a constant multiplied by a power of x, the slope of the log-log graph for y vs. x will give the power of x, and the y-intercept will be the log of the constant.

(Note: The properties listed here apply to both common logs and natural logs, as well as to logs of any other base.)

TRIGONOMETRY

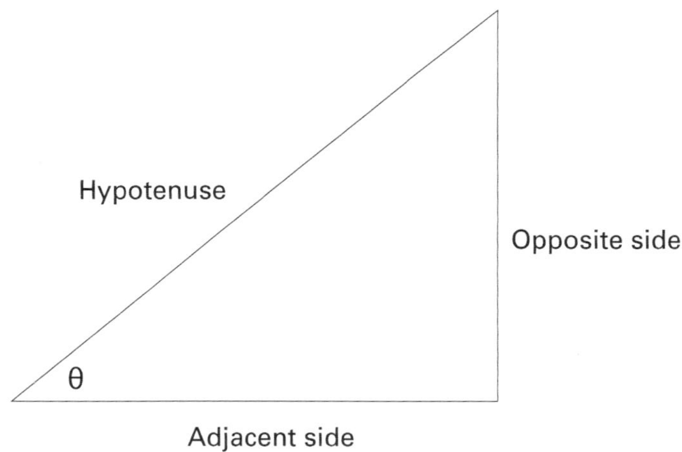

In a right triangle like the one shown above,

$$\sin \theta = \frac{opposite\ side}{hypotenuse}\ , \qquad \cos \theta = \frac{adjacent\ side}{hypotenuse}\ , \qquad \tan \theta = \frac{opposite\ side}{adjacent\ side}$$

Combining the above equations, it is easy to obtain,

$$\tan \theta = \frac{\sin \theta}{\cos \theta}$$

Sine, cosine and tangent are *periodic* functions, which means that they repeat in a regular pattern as the value of the argument (θ) changes. The sine and cosine functions are well behaved at all values of θ but the tangent function becomes undefined at values of $\theta = 90 \pm n$ (180°), where n is a positive integer. This means that the functions sine and cosine can be used to describe regular sinusoidal oscillations. The equation for these oscillations is given by:

$$y = A \cos (\omega t + \varphi)$$

where A is the amplitude of the oscillation, ω is the frequency of the oscillation and φ is the phase angle. Note that since $\sin (\theta + 90°) = \cos \theta$, either sine or cosine can be used in this equation and only the phase angle will change.

Below is a short list of trigonometric identities that will be useful for this lab. For a more complete list of identities (and the corresponding proofs), consult any of various mathematics texts.

$$\sin^2 \theta + \cos^2 \theta = 1$$

$$\sin (a + b) = \sin (a) \cos (b) + \cos (a) \sin (b)$$

$$\cos (a + b) = \cos (a) \cos (b) - \sin (a) \sin (b)$$

GRAPHING

The equation of a straight line is given by: $y = mx + b$ where m is the slope of the line and b is the y-intercept of the line. The slope of the line indicates the rate of increase of y with respect to the rate of increase of x. The slope is found as follows:

$$m = \frac{y_2 - y_1}{x_2 - x_1}$$

A slope of 1 indicates that x and y are changing at the same rate. The greater the slope, the faster y increases with respect to x. A line parallel to the x-axis has a slope of zero ($y_2 - y_1 = 0$), while one parallel to the y-axis has an infinite slope ($x_2 - x_1 = 0$). A positive slope indicates that the values of y increase as the values of x increase; a negative slope indicates that the values of y decrease as the values of x increase.

If you use a computer program to plot your graph and calculate the slope you should do a quick "sanity check" of the slope it gives you (especially if you use Excel). Just do a quick estimate of the slope from the above formula to make sure that the value the computer calculated is reasonable.

All graphs must have a title and axis labels. Each axis should be labeled with the quantity being plotted and its units. Axes must always be single-valued (no value repeats itself on the scale) and the scale must be linear (unless you are plotting a log-log graph using logarithmic graph paper or a graphing program on the computer, and then it is obviously logarithmic.) The graph must also be a full page in order to facilitate reading the data from it.

If you have a power-law equation the resulting graph will obviously not be a straight line if you use a linear scale. If the graph is not linear **do not** fit a straight line to it. If it appears to be a power-law you can do a power-law fit to it. (See page 21.)

A graph should also have error bars on all of the data points in both dimensions. These give you a measure of the uncertainty in the values being plotted.

Below is an example of a graph, done properly using the plotting program **Graphical Analysis** which is available on the computers in rooms 2010 and 2016.

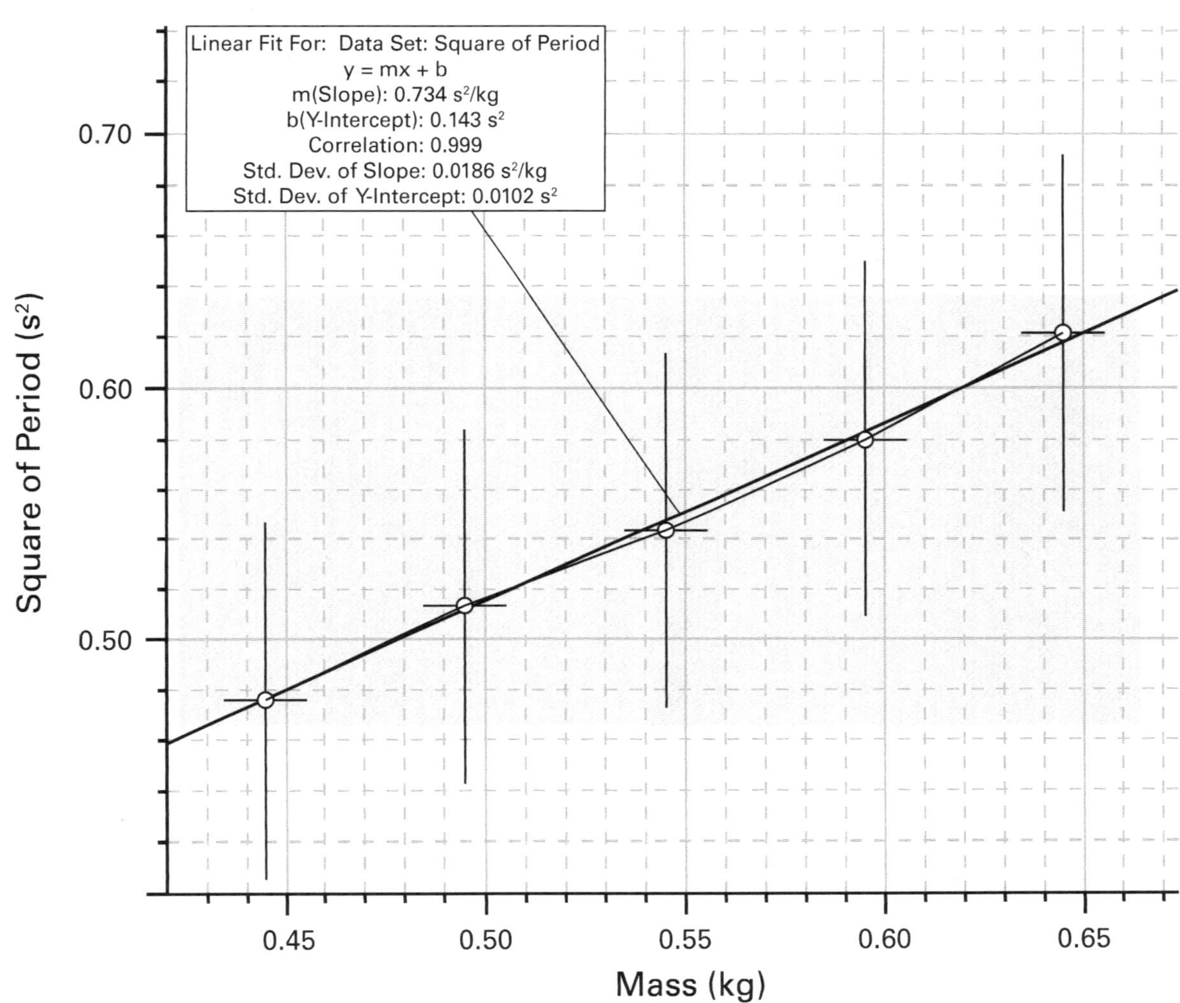

Summary of Important Tips on Graphing

1. On a y vs. x graph the independent variable (x) appears on the horizontal axis and the dependent variable (y) appears on the vertical axis.

2. All graphs should have a <u>title</u> and each axis should be labeled with <u>both</u> the <u>quantity</u> being graphed and the <u>units</u> in which it is measured.

3. Graphs should cover an <u>entire page</u>.

4. The slope should <u>always</u> be calculated from the graph, not from the data points.

5. If y is a linear function of x (that is, y depends on the first power of x), a least squares analysis can be used to obtain the best fit straight line. (**Excel** and **Graphical Analysis,** which are installed on the computers in Rooms 2010 and 2016, and also most other computer graphing programs include a least squares function.

6. If y is not a linear function of x (i.e., a graph of y vs. x does not produce a straight line), a power law should be tried. If y is a function of some power of x, a plot of log y vs. log x will produce a straight line, and the power can be obtained by the process described in the Logarithms section of the Math Review that precedes this section. (The graph can be obtained using **Excel, Graphical Analysis** or most other graphing programs, or by hand using log-log paper.)

7. If the dependence of y on x satisfies none of the above conditions, a smooth curve should be fitted to the data points. (**Graphical Analysis** will automatically "connect the dots" if you don't tell it to do anything else. This is not quite the desired smooth curve, but is the best option available in this case.)

Notes

Introduction:

- Accuracy = how close you are to the actual value
  - ↳ use % error p.13 $\%\,error = \dfrac{|experimental - theoretical|}{theoretical}\,(100\%)$

- Precision = how close the experimental values are to each other when the experiment is run several times
  - ↳ use % uncertainty $\%\,uncertainty = \dfrac{uncertainty}{experimental}\,(100\%)$

- If both precision and accuracy are about the same and small = good results
  - – Want percent error to be smaller than percent uncertainty (actual answer is w/in range of uncertainty)

- Uncertainty p.11
  - ↳ ① If trials are > 50, use standard deviation as uncertainty
  - ② If trials are 10-50, use average of deviations
  - ③ Fewer than 10 trials, use largest deviation
  - ④ Just one or two trials, make reasonable estimate

trials	deviation
9.9	$|10.5 - 9.9| =$
10.1	$|10.5 - 60.6| =$
11.0	$|10.5 - 11.0| =$
10.5	$|10.5 - 10.5| = 0$
⋮	⋮
10.5	$10.5 - 10.5 =$
avg. 10.5	

- Propagation of Uncertainty
  - ↳ Given two numbers w/ uncertainties

$$y \pm \delta y \qquad 10 \pm 1$$
$$z \pm \delta z \qquad 15 \pm 2$$

$$x = y + z$$
$$x = 10 + 15 = 25$$
$$\delta x = \delta y + \delta z = 1 + 2$$
$$\delta x = 3 \qquad x = 25 \pm 3$$

ⓐ Addition & Subtraction
  ↳ What is the uncertainty in x, if x is = to $y + z$

$$\delta x = \delta y + \delta z$$

ⓑ Multiplication & Division
  ↳ What is the uncertainty in x, if $x = yz$?

$$\frac{\delta x}{x} = \frac{\delta y}{y} + \frac{\delta z}{z}$$

$$\frac{\delta x}{150} = \frac{1}{10} + \frac{2}{15} = 0.23 \longrightarrow \text{fractional uncertainty of } x$$
$$\Rightarrow 20\% \quad \%\,uncertainty$$

$$\delta x = 35$$

↳ $x = \dfrac{y}{z}$

$$\frac{\delta x}{x} = \frac{\delta y}{y} = \frac{\delta z}{z} = 0.23$$

$$\frac{\delta x}{0.67} = 0.23$$

$$= .156$$

$$x = .67 \pm 0.16$$
$$\text{or}$$
$$0.7 \pm 0.2$$

24

EXPERIMENT 1

Nominal Resistance and the Multimeter

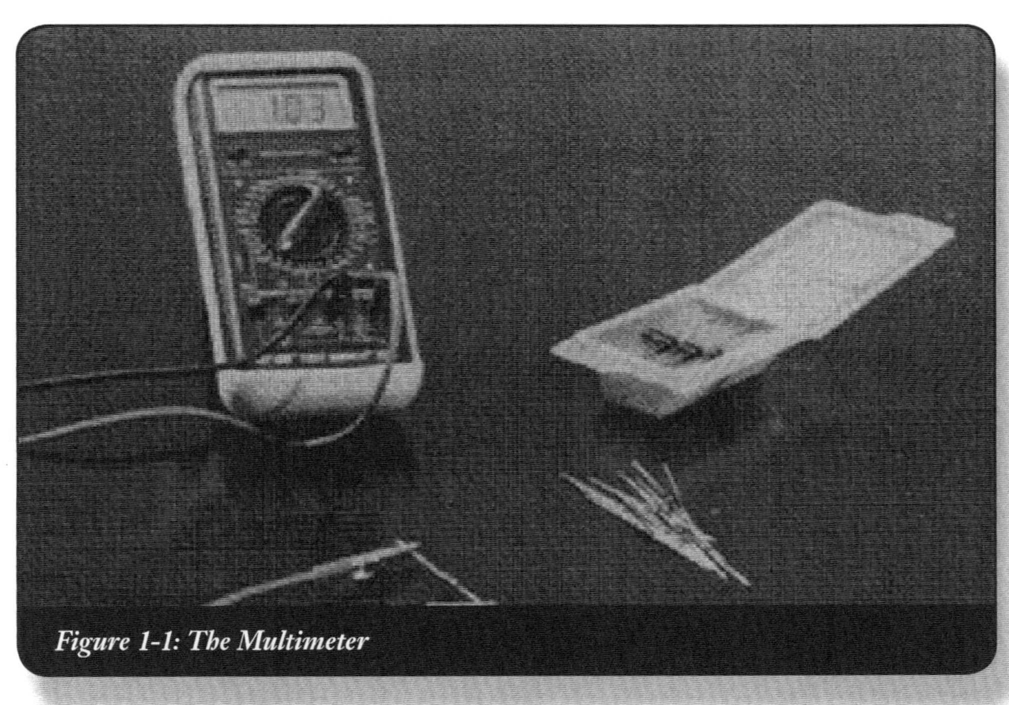

Figure 1-1: The Multimeter

INTRODUCTION

Commercial resistors come in a wide variety of composition, packaging, resistance values, power tolerances, and precision. In this experiment you will use a multimeter to measure the value of 100 resistors nominally of the same value, plot a histogram of your results, and compare the average resistance with the nominal value and the deviation with the nominal precision. Nominal values are indicated on the body of the resistor in a code of colored bands; a chart illustrating this code is posted in the laboratory.

PROCEDURE

P1. Using the color code (Figure 1-2), determine the nominal value of the resistors you have been given. Start with the stripe closest to one end and read the colors in order. The first two stripes give the first and second digit of the resistance value respectively, and the third stripe gives the power of ten to which the number thus obtained should be multiplied. For example, if the first three stripes are blue, yellow and brown, then the nominal resistance is $64 \times 10^1 \ \Omega = 640 \ \Omega$.

P2. Examine the multimeter and familiarize yourself with the settings needed to measure various quantities. You will be using a similar meter in other experiments in this laboratory to measure not only resistance, but capacitance, voltage, and current. (See Figure 1-3.) For this experiment, choose an appropriate setting to measure resistance in ohms (Ω) or kilo-ohms ($k\Omega$), based on your result in P1. If you have difficulty with this, ask your instructor for help.

P3. Holding the red probe on one of the two resistor leads and the black probe on the other, read the resistance value from the display.

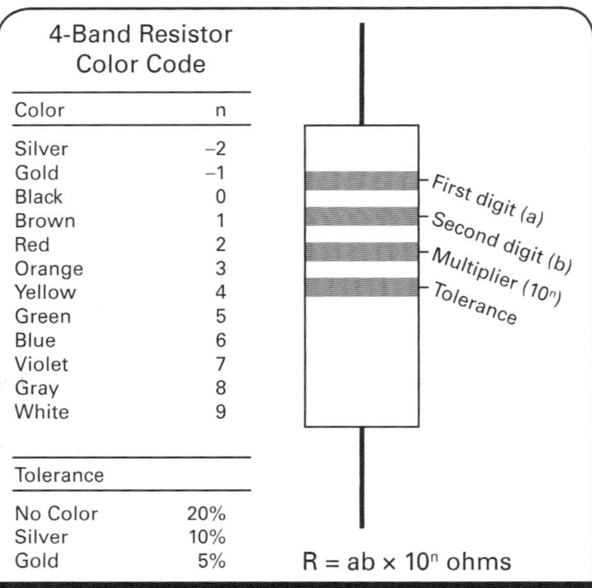

Figure 1-2: Reading the Resistor Code

Figure 1-3: Various Types of Multimeters

CALCULATIONS

C1. From the data of P2 calculate the average resistance and the standard (root-mean-squared) deviation from the average. Compare these values with the nominal values obtained from the color code. Are all the measured resistance values within the stated tolerance?

C2. Construct a histogram representing the data obtained in P2. If necessary, review the section "The Mathematics: Distribution in Random Error" starting on page 9 in this manual and Experiment 1: "Measurement Precision and Distribution" in *The Physics Lab Manual: Part 1* (pages 25–30).

C3. Recall that in a normal distribution 68% of the values obtained should lie within one standard deviation of the average, 95.5% within two standard deviations, and 99.7% within three standard deviations. Compare your results with these predictions.

Brown, Green, Brown, Gold

Trial #

1	148.3
2	149.2

Coulomb's Law $\quad F = \dfrac{kQ_1Q_2}{r^2} = \dfrac{kQ^2}{r^2}$

Instead of measuring F, we measure θ ($\theta \propto F_1$)

Instead of measuring Q (charge) we measure V (voltage) $(V \propto Q)$

$$\theta \propto \dfrac{V^2}{r^2}$$

$\log \theta$ vs. $\log r$
    Expected slope = -2

$\log \theta$ vs. $\log V$
    expected slope = 2

$6 \, kV$

r			
14 cm	33°	32°	33°
10 cm	55°	60°	57°
9 cm	67°	71°	73°
8 cm	79°	83°	82°
7 cm	106°	106°	110°
5 cm	165°	172°	169°

7 cm

V			
5 kV	85°	84°	84°
4 kV	49°	51°	55°
3 kV	35°	33°	34°
2 kV	18°	15°	17°
1 kV	9°	9°	8°

Coulomb's Law

Figure 2-1: The Coulomb Torsion Balance and High Voltage Charging Supply

INTRODUCTION

A common U.S. dime has a mass of just over two grams; if one could take two grams of protons and hold them one meter from two grams of electrons the resulting force would equal 6×10^{23} Newtons. To observe a gravitational force of roughly the same magnitude, one could try weighing the Moon (at the Earth's surface!). Because electrical forces are so large it is nearly impossible to separate even small amounts of positive from negative charges, and laboratory efforts to investigate Coulomb's law must rely on sustaining relatively small charge differentials; as a result, the forces which can be investigated are very small, and sensitive apparatus must be used. In this experiment the force between two charged spheres is investigated using a sensitive torsion balance, or pendulum, similar to the balance used by Coulomb himself, and by Cavendish at about the same time to determine the Universal Gravitational Constant G (see Figure 2-1).

In principle the method is simple: two spheres are electrically charged from a high-voltage power supply, the spheres held a known distance apart, and the resulting force is determined by measuring the angular twist produced in the wire by which one sphere is suspended. Varying the distance while measuring the corresponding force determines the distance dependence in the force law, and varying the amount of charge on each sphere (by varying the output voltage of the power supply used to charge the spheres) determines the charge dependence. It is also possible to use this same apparatus to obtain a numerical value for the constant of proportionality appearing in Coulomb's law, but this requires more time than one laboratory session permits, and will not be attempted.

Read these instructions through carefully and completely while examining the apparatus. The torsion wire can easily be broken and must be treated with care. To obtain good results from this equipment the experiment must be approached deliberately and executed with patience. In particular, pay attention to the following points:

1. Atmospheric humidity can be a source of charge leakage, and if the humidity is high you may experience frustration in obtaining results. In such a case, try drying the apparatus using a small hair dryer immediately before charging the spheres, and take measurements as soon as possible thereafter.

2. Dry atmospheric conditions can also cause problems; charge will build up on clothing, walls, etc., with no way to leak to ground. To avoid this, don't do the experiment if you are wearing synthetic fabrics, keep the apparatus away from walls or other objects which could be charged, stand directly behind the apparatus at a maximum comfortable distance from it while taking measurements, and if static conditions are extreme connect yourself to a grounding wire.

3. Eliminate room drafts as much as possible; use cardboard baffling if necessary.

4. Avoid handling the spheres and the rods which support them; surface contamination can cause charge leakage. Use alcohol to clean the surfaces if they have been handled.

5. To minimize leakage effects, take measurements as quickly as possible after charging the spheres.

Patience and perseverance will pay dividends in this very interesting examination of one of the fundamental laws of nature. You will be surprised at the strength of the effect produced by changing the distance between the charged spheres, and will acquire a good intuitive feeling for an inverse-square force law if you do the work carefully.

PROCEDURE

P1. Your instructor will have performed the initial adjustments of the apparatus, but you should check to see that this has been done properly. Compare Figures 2-1, 2-2, and 2-3 with the apparatus, and identify the components before making any adjustments.

Note that the index arm which serves to mark the reference position of the counterweight vane has been clamped to the vane by means of a thumbscrew (Figure 2-4). Carefully loosen the thumbscrew and rotate the index arm away from the counterweight vane, leaving the index arm parallel to the mark on the rear length of the apparatus. The pendulum should hang freely between the magnets of the damping arm and level with the base (the small springs clipped to the vane provide a means of adjustment to help achieve balance), the degree scale on the torsion knob at the top of the apparatus should read zero, and the index line on the vane should align with the mark on the index arm (Figure 2-5). In addition the two spheres should be at the same height and aligned laterally. If any of these adjustments are incorrect, ask your instructor to realign the apparatus.

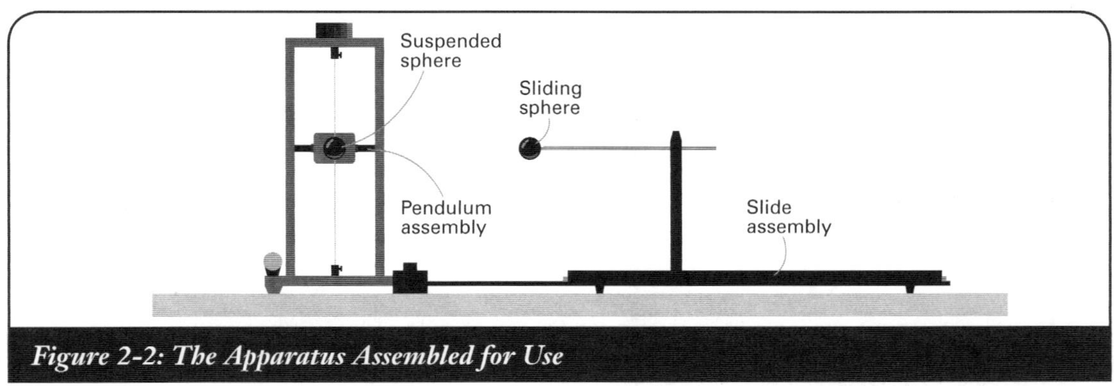

Figure 2-2: The Apparatus Assembled for Use

P2. Note carefully how the high-voltage power supply is connected (Figures 2-1 and 2-6): The probe used to charge the spheres is connected to the +6 kV (far-right) terminal, and the far-left terminal has been connected to a ground tap on the power strip mounted on the end of the laboratory bench. Make certain this ground connection is secure. Do not turn on the power supply until you are ready to charge a sphere, and turn the supply off immediately after charging; high voltage at the terminals can cause leakage currents which will affect the balance. The amount of charge transferred to a sphere is determined by the voltage delivered by the supply, and is varied with the **HIGH VOLTAGE ADJUST** knob.

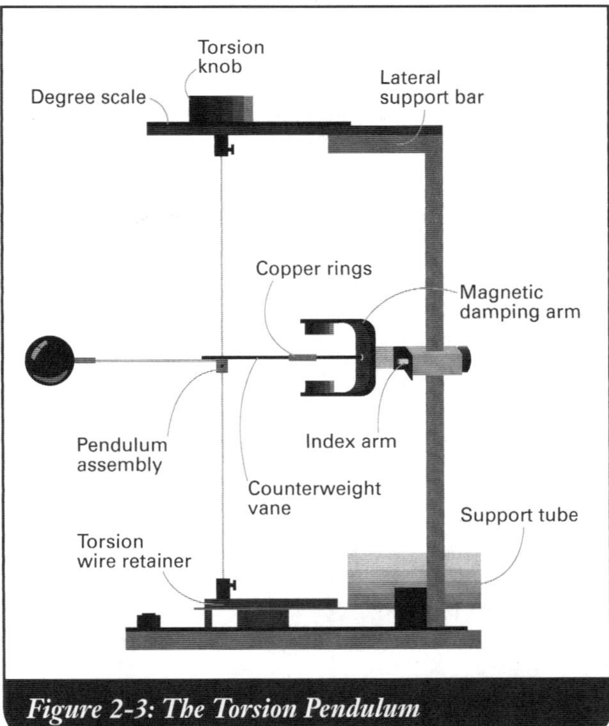

Figure 2-3: The Torsion Pendulum

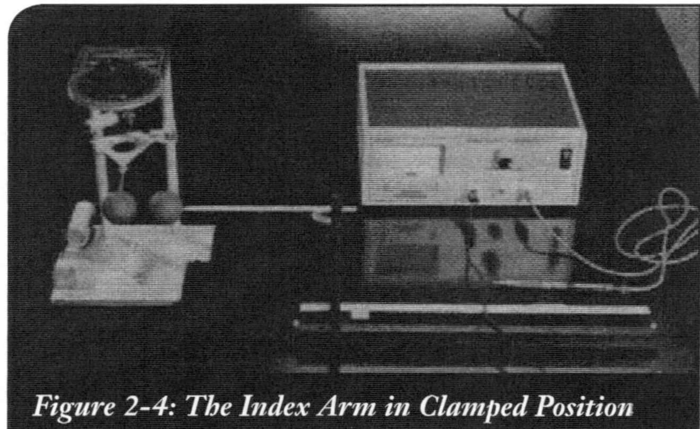

Figure 2-4: The Index Arm in Clamped Position

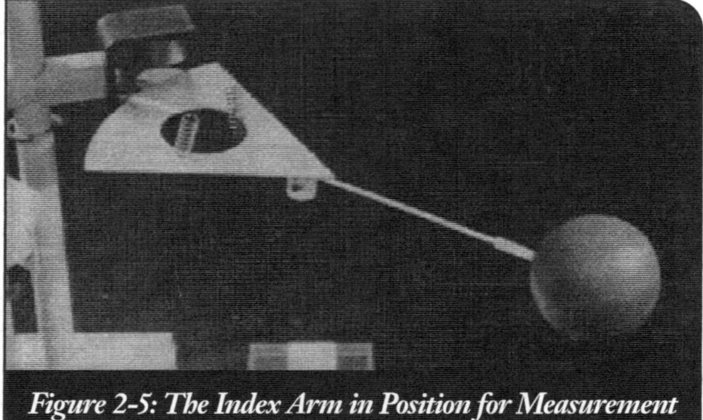

Figure 2-5: The Index Arm in Position for Measurement

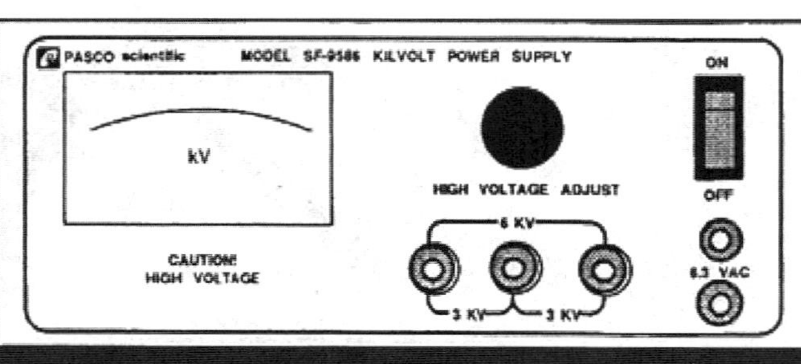

Figure 2-6: The Kilovolt Power Supply

NOTE!

This power supply has been current limited (2 mA at lower voltages, 0.1 mA at the full voltage output) for safety, but high voltage supplies should always be used with caution. Indeed, any voltage greater than about 35 volts should be considered potentially dangerous since the current produced is determined not only by the voltage but by the resistance across which the voltage is applied, and it is current which causes biological damage.

Do not touch the probe when the supply is on, and leave the supply on only momentarily while charging a sphere. Isolate yourself from ground to the extent possible while using the supply.

P3. Begin measurements by making certain that the spheres are initially discharged: touch them with a grounding probe (a wire connected to a ground tap on the laboratory bench power strip). Slide the sliding sphere to the far end of its track, as far as possible from the suspended sphere. Check that the torsion dial is set to zero, and zero the balance by gently rotating the small retainer at the bottom of the torsion wire until the pendulum assembly is at its zero displacement position as indicated by the index marks on the counterweight vane and index arm.

P4. With the spheres at maximum separation, use the charging probe to charge both spheres to 6 kV. Immediately after charging the spheres, turn the power supply off.

P5. Position the sliding sphere at a position of 20 cm.[3] Adjust the torsion knob as necessary to balance the forces and bring the pendulum back to the zero position. Record the distance R and the angle θ.

P6. Repeat steps P3–P5 several times, until your results are consistent to within one degree.

P7. Repeat steps P3–P6 for distances of 14, 10, 9, 8, 7, and 5 cm. These data will enable a determination of the distance dependence in Coulomb's law.

P8. Again make certain both spheres are discharged. Then with the spheres at maximum separation, charge both spheres to 6 kV and move the sliding sphere to a convenient position that allows for a good measurement. Record the force (i.e., the angle θ as in P5).

Note

[3] Notice that if the two spheres are in contact the scale reads 3.8 cm, which is the diameter of each sphere; hence the scale reading at any other position is the center-to-center separation of the spheres.

P9. Again discharge the spheres and separate them to maximum distance. Then recharge both spheres to 5 kV and move the sliding sphere to the same position as in P8. Record the force as before.

P10. Repeat P9 for charges of 4 kV, 3 kV, 2 kV, and 1 kV. Remember to repeat each measurement until a consistent result is obtained for a given voltage.

P11. When you have finished all your measurements make certain the spheres are discharged and the power supply is off, and reclamp the index arm to the counterweight vane. Make certain there is no stress on the wire.

CALCULATIONS

C1. The restoring force exerted by the torsion wire is directly proportional to the angle of twist, so to investigate the distance dependence in Coulomb's law it is only necessary to look for a relation between the angles recorded in steps P3–P7 and the corresponding separations between the spheres. Assuming the relation to be a power law, the slope of the graph of $\log \theta$ vs. $\log R$ should give the power of R on which θ and thus the force (F) depends. Using your data from P5–P7, plot this graph and obtain a linear fit. Calculate the slope and its uncertainty. Is your result consistent with Coulomb's law?

$$F = k \; \frac{Q_1 Q_2}{R^2}$$

C2. Next plot θ vs. R directly, and try to fit the curve to a second-order polynominal. You will find that the inverse square law appears to hold for larger values of R but not for smaller. This is because the spheres used are not simple point charges, and when the two spheres are near one another they affect each other's charge distribution, making the distributions non-spherical with the effective centers of the distributions at some points other than the center of each sphere.[4]

C3. Using the data obtained in P8–P10, plot θ vs. V and $\log \theta$ vs. $\log V$. Again it is reasonable to suspect a power law, so again use the slope of your log graph to determine the power of V and thus the power of the charge (Q) on which the angle (θ) and thus the force (F) depends. (Do a linear fit for the log graph and calculate the slope and its uncertainty as before.) Again comment on whether your results are consistent with Coulomb's law.

Note

[4] A correction factor can be introduced into the calculation for this effect: If each value of Q is multiplied by the quantity $(1 - 4r^3/R^3)$, where r is the radius of each sphere and R is the separation between spheres, a plot of the corrected Q vs. R will more closely fit the inverse-square relationship.

$$F = \frac{K Q_1 Q_2}{r^2} \qquad \varepsilon_0 = \frac{1}{4\pi K}$$

$$F = \frac{Q_1 Q_2}{4\pi \varepsilon_0 r^2}$$

Capacitor: $C = \frac{\varepsilon_0 A}{d}$

energy of capac. $U = \frac{1}{2} C V^2$

$$W = Fd = U$$

$$Fd = \frac{1}{2} C V^2$$

$$Fd = \frac{1}{2} \frac{\varepsilon_0 A V^2}{d}$$

$$F = \frac{\varepsilon_0 A V^2}{2 d^2}$$

Use the slope of F vs. V^2 to find ε_0

$$\text{Slope} = \frac{\varepsilon_0 A}{2 d^2}$$

EXPERIMENT 3

The Electrostatic Balance and the Permittivity of Free Space

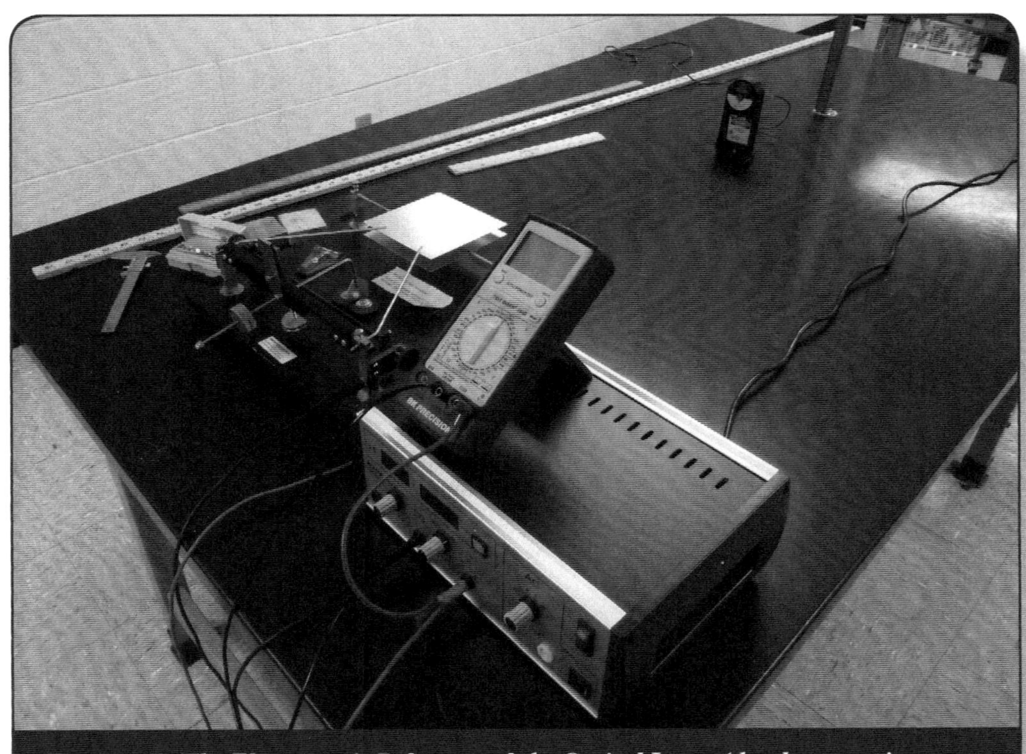

Figure 3-1: The Electrostatic Balance and the Optical Lever (the electrostatic balance is on the right, the laser and meter stick for the optical lever are on the left)

INTRODUCTION

In the previous experiment an investigation was made of the functional relationships described by Coulomb's Law. That is, the magnitude of the electrostatic force between two charged objects is directly proportional to the product of the magnitudes of the two charges (q_1 and q_2) and inversely proportional to the square of the distance between them (r_{12}).

$$F_{12} = \frac{q_1 q_2}{r_{12}^2} \qquad (1)$$

CAUTION!

DO NOT TOUCH THE EXPERIMENTAL APPARATUS UNTIL YOU HAVE READ THE INSTRUCTIONS.

The apparatus incorporates delicate knife edges which can be damaged easily. Please treat the apparatus with care. Do not look directly into the laser beam; to do so could cause severe eye damage.

TAKE NO CHANCES.

If all quantities in this equation are expressed in SI units, the proportionality constant is the Coulomb constant k, or alternatively $1/(4 \doteq \varepsilon_0)$, where ε_0 is the permittivity of free space, which appears in Maxwell's equations. It is the constant ε_0 which we seek to obtain here. (Note that $\varepsilon_0 = 1/(4 \pi k)$.) Thus Equation (1) becomes

$$F_{12} = \frac{1}{4 \pi \varepsilon_0} \frac{q_1 q_2}{r_{12}^2} \qquad (2)$$

CALCULATING THE PERMITTIVITY OF FREE SPACE

The energy stored in the electric field between the two plates of a parallel plate capacitor with capacitance C and potential difference V between the plates is[5]

$$U = \frac{1}{2} C V^2 \qquad (3)$$

If the plates are separated by a distance d, this energy can be thought of as the work W done in moving one plate a distance d away from the other plate against the attractive force F between the two plates. (Note that the force between the plates actually varies as the plates are moved with respect to one another. The force F here can be thought of as the average force between the plates during this operation.)

$$W = F d \qquad (4)$$

Setting $U = W$ in these equations, we obtain

$$F d = \frac{1}{2} C V^2 \qquad (5)$$

Note

[5] See for example Giancoli, *Physics, Principles with Applications* (5th ed.) p. 517, or Wolfson, Pasachoff, *Physics with Modern Physics for Scientists and Engineers* (3rd ed.), p. 669.

For a parallel plate capacitor where each of the plates has area **A** and they are separated by the distance **d,** the capacitance is given by[6]

$$C = \frac{\varepsilon_0 A}{d} \qquad (6)$$

Substituting (6) into (5), we obtain

$$F d = \frac{1}{2}\left(\frac{\varepsilon_0 A}{d}\right) V^2 \qquad (7)$$

or, rearranging,

$$F = \frac{\varepsilon_0 A V^2}{2 d^2} \qquad (8)$$

If we place several small masses, each of mass **m,** on the top plate (with the voltage turned off) and note the position of the upper plate, then take one of the masses off and turn on the voltage and increase it until the upper plate comes to the same position as it did with that mass on it, then the electrostatic force in (8) will be the same as the gravitational force on the mass **m.** That is, **F = mg.** So we have

$$m g = \frac{\varepsilon_0 A V^2}{2 d^2} \qquad (9)$$

If this procedure is repeated, removing **2m, 3m,** etc., keeping the spacing **d** between the plates constant, a graph of **F** vs. **V²** can be obtained. The slope of that graph will be

$$slope = \frac{\varepsilon_0 A}{2 d^2} \qquad (10)$$

If the area **A** and the separation **d** are measured for the plates described, this slope can be used to obtain an experimental value for ε_0:

$$\varepsilon_0 = \left(\frac{2 d^2}{A}\right) slope \qquad (11)$$

Note

[6] Giancoli, *Physics, Principles with Applications* (5th ed.) p. 513, or Wolfson, Pasachoff, *Physics with Modern Physics for Scientists and Engineers* (3rd ed.), p. 667.

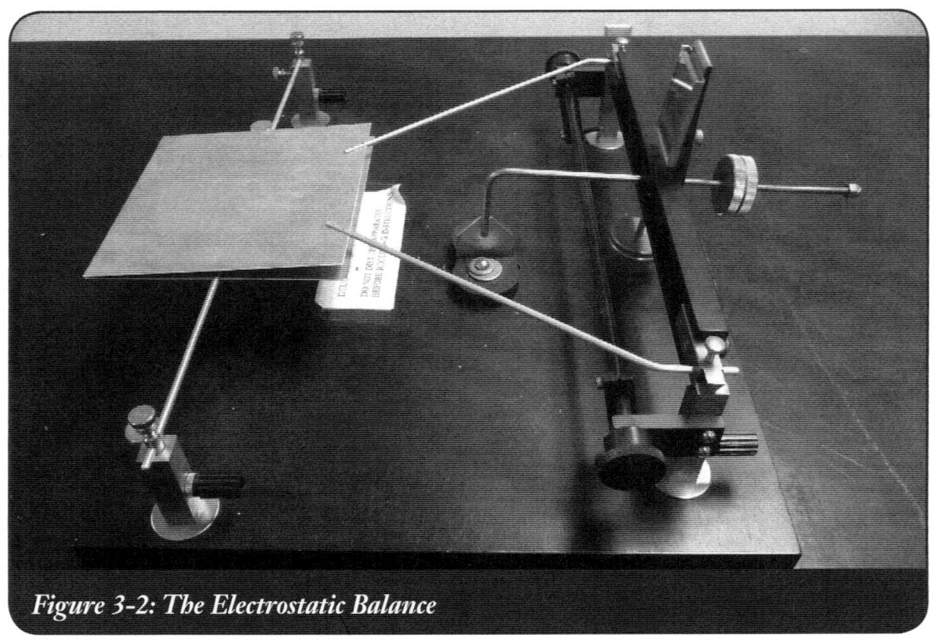

Figure 3-2: The Electrostatic Balance

USE OF THE LASER AS AN OPTICAL LEVER TO MEASURE SMALL DISTANCES

The distance, **d,** between the plates described above cannot be conveniently measured in a direct manner, so the laser is used to make an indirect measurement. If the plates start out with no spacing between them and then the top plate is moved slightly upward, it moves through an angle θ, as shown in Figure 3-3. The distance **a** in the diagram is the horizontal distance from the mirror to the center of the fixed plate.

As the mirror moves through an angle θ, the incident angle of the laser on the mirror also moves through the angle θ. Since the reflected angle is equal to the incident angle, it also increases by the angle θ. So as the mirror moves through angle θ, the laser beam reflected from

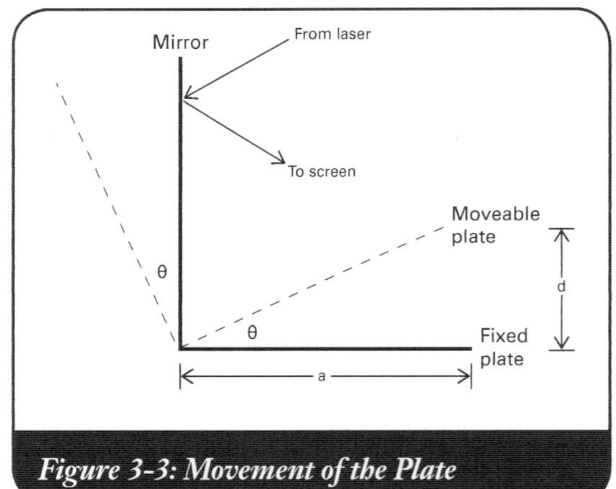

Figure 3-3: Movement of the Plate

the mirror moves through an angle of 2θ. Figure 3-4 (a) shows the triangle representing the movement of the plate and Figure 3-4 (b) shows the triangle representing the movement of the reflected laser beam. The vertical leg of this second triangle is the distance the laser beam moves along the meter stick as the plate moves, so the horizontal leg is the distance from the mirror to the meter stick (**b**).

If the displacement of the plates is sufficiently small, both θ and 2θ satisfy the small angle approximation, i.e. tan θ ≈ θ. From the triangle in Figure 3-4 (a), we see that tan θ = **d/a**, and from the triangle in Figure 3-4 (b), tan 2θ = **D/b**. So θ ≈ **d/a**, and 2θ ≈ **D/b.** Combining these, we obtain

$$\frac{D}{b} = \frac{2\,d}{a} \qquad\qquad (12)$$

or

$$d = \frac{D\,a}{2\,b} \qquad\qquad (13)$$

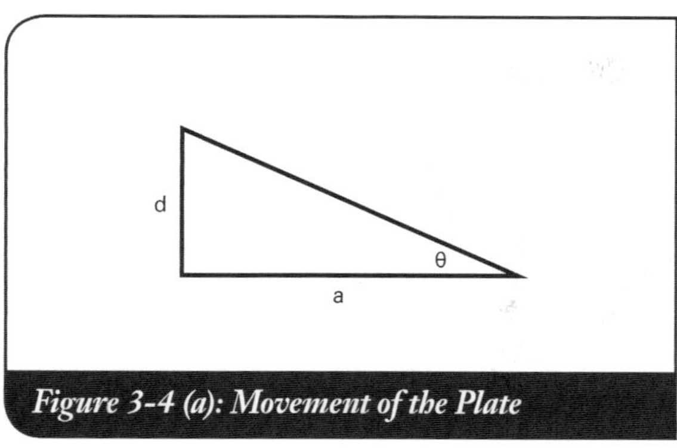

Figure 3-4 (a): Movement of the Plate

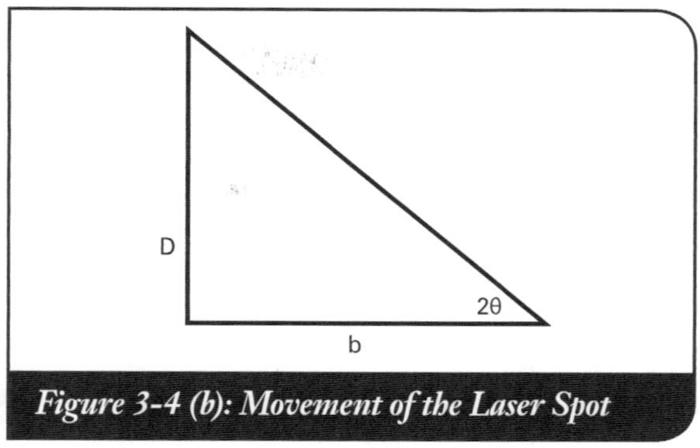

Figure 3-4 (b): Movement of the Laser Spot

Equation (13) can be used to calculate the distance, **d**, between the plates if the quantities **a, b,** and **D** are measured.

PROCEDURE

P1. Carefully inspect the equipment and compare it to Figures 3-1 and 3-2. Small supporting blocks are inserted under the moveable frame to hold it off the knife edges. Do not remove these blocks until you understand the beam-rest mechanism used to keep the knife edges from resting on the posts: adjacent to the knife edges on the under side of the moveable frame are two small indentations or dimples. Into these may be rotated two pointed brass pegs by turning the knurled knobs mounted on supporting posts attached to the base (Figure 3-2). With the pegs inserted into the dimples the moveable frame can be lifted onto or off the knife edges. The frame should be supported off the knife edges until a measurement is to be taken. Begin by using the spirit level to level the base of the balance, making certain the base is resting firmly on the table; small vibrations, even air drafts, will contribute error in the result.

P2. Measure and record the length and width of the plates. (Measure both plates to see if there is any difference between them and use the average values for the length and the width.)

P3. When you have familiarized yourself with the equipment and are ready to take measurements, follow the precautions given in P1 while using the beam-lift knobs to carefully lift the frame and remove the wood blocks. When properly mounted the vane attached to the moveable frame will swing freely between the two damping magnets on the base of the balance.

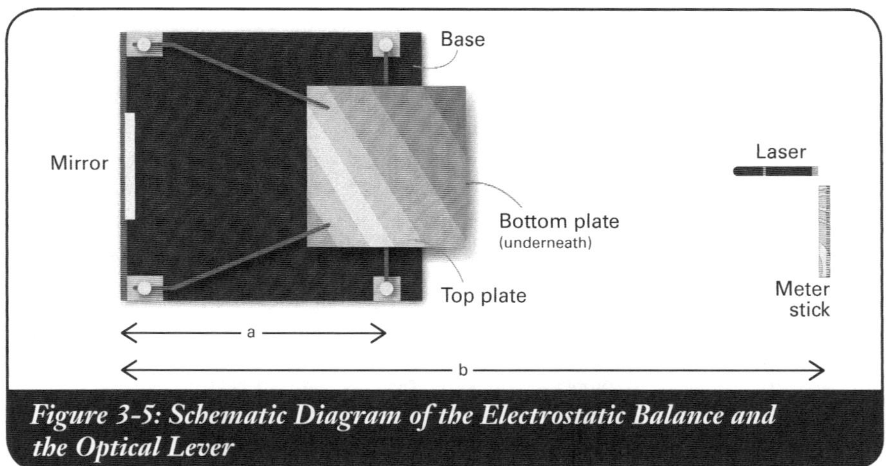

Figure 3-5: Schematic Diagram of the Electrostatic Balance and the Optical Lever

P4. Place enough weight on the top plate to bring the plates together so that the spacing, **d,** between them is zero. (Use a coin or a large weight from the weight set provided.) Turn on the laser and aim it so that it reflects from the mirror and hits the meter stick, which is positioned vertically at a horizontal distance **b** from the mirror. Record the position of the laser spot on the meter stick as h_0.

P5. Measure and record the distances **a** (the horizontal distance from the mirror to the center of the fixed plate) and **b** (the horizontal distance from the mirror to the meter stick) as shown in Figure 3-5.

P6. Now remove the heavy weight and place five small weights (less than 20 mg each) on the top plate so that the plates **do not** come together (adjust the counterweights if necessary). Record the new position of the laser spot on the meter stick as **h** (be careful **not** to move the apparatus once this measurement is made).

P7. Next remove one of the small weights and connect the power supply to the plates (i.e., connect the positive terminal to one plate and the negative terminal to the other). Turn up the voltage until the laser spot appears on the meter stick in the same position (**h**) as it did in P6. Record this voltage.

P8. Now remove another small weight from the upper plate and readjust the voltage until the laser spot is again at **h,** as in P6. Record the total amount of mass removed (**2m**) and the voltage, **V.**

Note that h remains <u>constant</u> for <u>all</u> masses and voltages.

P9. Repeat P8, removing another of the small masses each time until there are no masses left. Be sure to record the voltage needed to bring the spot back to **h** when there are no masses on the upper plate.

P10. Make a table showing the values of **m** and **V.** (Remember that **m** is the amount of mass **<u>removed</u>.**)

CALCULATIONS

C1. Calculate $D = h - h_0$. This quantity will be used to calculate the distance, **d,** between the plates (see above).

C2. Use the length and width measurements from P2 to calculate the area **A** of the plates.

C3. Calculate the spacing, **d,** between the plates by substituting **a** and **b** from P5 and **D** from C1 into Equation (13).

C4. Make graphs of gravitational force **F** vs. **V** and log **F** vs. log **V** for the range of masses used. Use these graphs to investigate the power dependence of **F** on **V.** (Use a best fit straight line for the log graph, but **not** for the **F** vs. **V** graph.) Compare your result with the power dependence predicted by Equation (9).

C5. Make a graph of **F** vs. $\mathbf{V}^2$ and use the slope of this graph along with the information from C2 and C3 to calculate ε_0 from Equation (11). Compare your experimental value to the "book" value:

$$\varepsilon_0 = 8.85 \times 10^{-12} \ \mathrm{C^2/N \ m^2}$$

Notes

$T = \frac{1}{f}$

$f = \frac{1}{T}$

EXPERIMENT 4

The Oscilloscope

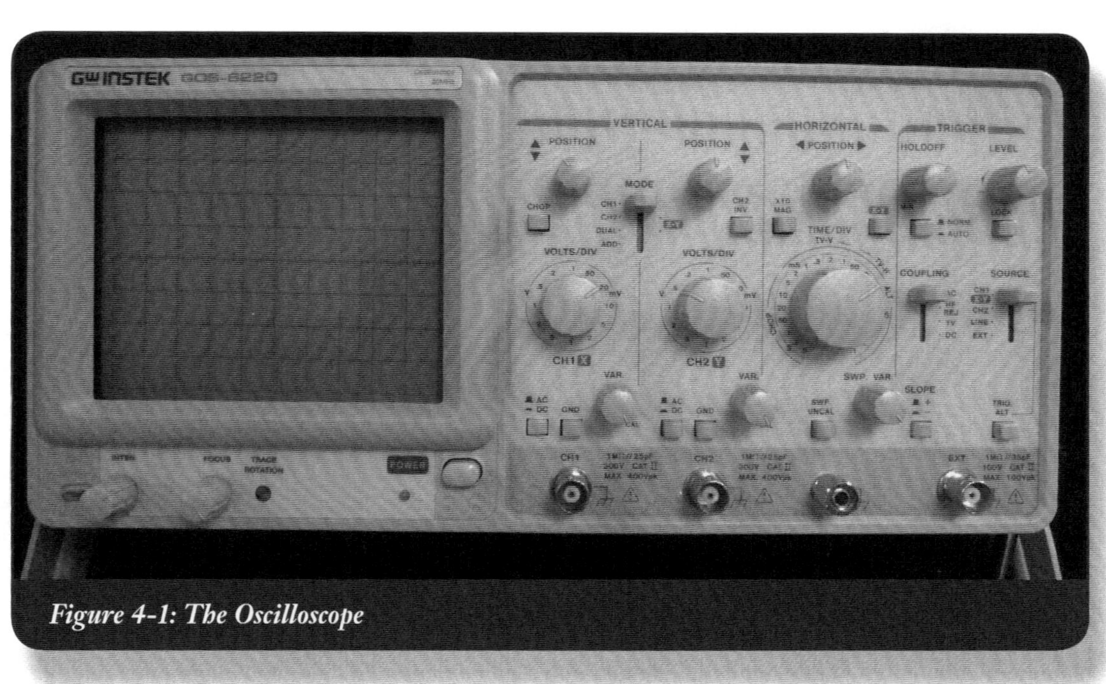

Figure 4-1: The Oscilloscope

INTRODUCTION

This experiment will explore the functions and controls of a *dual-channel* oscilloscope, a device capable of receiving input from two different voltage sources and outputting two displays of voltage over time simultaneously. Not all oscilloscopes will have the same capabilities, but all will be able to provide a display of voltage over time. This experiment will provide a basic understanding of an oscilloscope's controls and functions in a concise manner. If more detailed information is required, consult the Operator's Manual supplied with the oscilloscope.

PRELIMINARIES

The oscilloscope uses *BNC connectors*[7] and coaxial cables. For input sources that also use a BNC connector, the connection can be made directly, using a coaxial cable with BNC connectors on each end, like the one shown in Figure 4-2. For sources that use standard leads, an adapter like the one shown in Figure 4-3 is used. If you wish to connect two coaxial cables to the same terminal on an oscilloscope or other device that uses BNC connectors, a *t*-connector, (Figure 4-4) is used.

Connections across small circuit elements such as resistors are sometimes made using a probe (Figure 4-5). The probe has a spring loaded tip that is connected to the positive terminal of the element and an attached wire (black) that is used for the negative terminal. Figure 4-6 shows how a probe might be used to measure voltage across a resistor.

Some probes will have a slide switch with a *×1* and a *×10* setting. At *×1* the probe will apply the full measured voltage to the scope terminal, but at *×10* will apply only one-tenth of the measured voltage; i.e., the probe provides a ten-to-one attenuation of the applied voltage. Other probes may have a *×10* attenuator fixed next to the BNC connector.

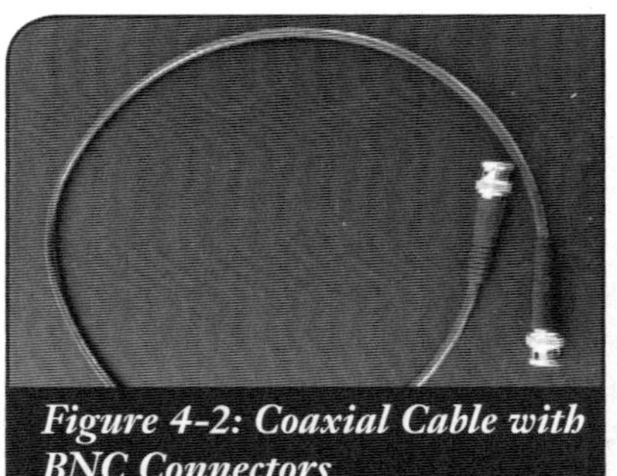

Figure 4-2: Coaxial Cable with BNC Connectors

Figure 4-3: BNC to Standard Lead Adapter

Figure 4-4: t-Connector

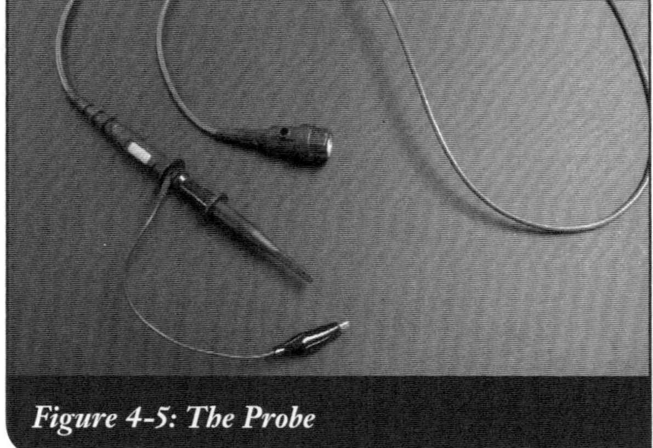

Figure 4-5: The Probe

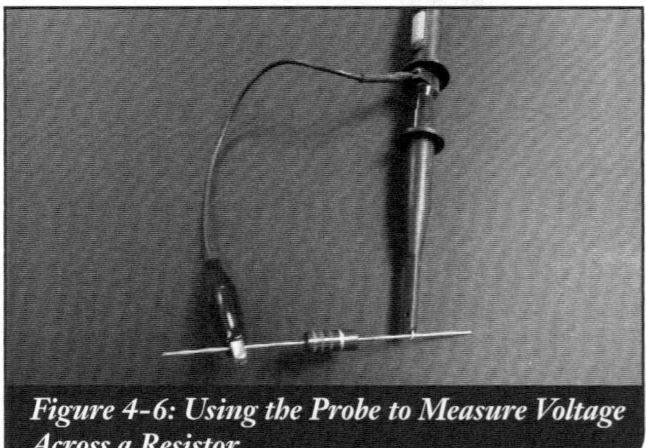

Figure 4-6: Using the Probe to Measure Voltage Across a Resistor

Note

[7] BNC stands for *bayonet coupling*. Coaxial cables must be matched carefully to their connectors in order to provide an uninterrupted path for the signal. Connectors are identified by codes related to their construction and size. Other examples are MHV (BNC modified for high voltage), TNC, SSMA, UHF, etc.

PROCEDURE

P1. Locate the POWER button and turn on the oscilloscope. A green or blue horizontal line, the *baseline trace*, should appear on the screen. The oscilloscope may need a minute or two to warm up, so be patient. If a few minutes have passed and the screen is still blank or displays only a dot instead of a line, ask your instructor for help.

Once the trace appears, it may look dim or out of focus. The INTENSITY and the FOCUS knobs can be used to produce a clear image. Adjust both knobs until the line is sharp and narrow and comfortably bright. However, keep in mind that setting the intensity too high may damage the screen.

P2. The oscilloscope's screen is divided into a grid by several vertical and horizontal lines. These lines represent divisions and create a graph on which the horizontal axis represents time and the vertical axis represents voltage.

You will notice three large knobs on the front of the oscilloscope. Two of these are identical and control the voltage scales for channels 1 and 2. (We will use only one for now.) The position of each of these knobs determines how many volts are represented by one vertical division on the screen. For example, if the knob is set at five volts per division, then each vertical division will represent five volts. Each division is further divided into smaller portions, representing fractions of a division. If a division is divided into five subdivisions, then each subdivision represents ⅕ of a division. This means that if the voltage scale is set at *5 volts/div*, then each subdivision represents 1 volt. But if the voltage setting is *1 volt/div*, then each subdivision represents 0.2 volts.

Below each voltage knob is a switch labeled AC-GND-DC. The position of this switch should match the input, AC or DC, being used for that channel. If the switch is set to GND, that channel will be grounded and display a voltage of zero regardless of the input.

DC VOLTAGE MEASUREMENTS

Find the switch labeled VERTICAL MODE and CH 1-CH 2-BOTH. This switch determines whether the oscilloscope displays the input from channel 1, channel 2 or both. Make sure the VERTICAL MODE switch is set to CH 1 and channel 1 is set to receive a DC input.

There should be two knobs above the voltage knobs, each labeled POSITION. Twist these back and forth to see that one knob controls the vertical position and one controls the horizontal position. Adjust the horizontal position so that the green or blue line is centered on the screen. The vertical position knob can then be used to set the "zero level" for DC voltage measurements. Set the vertical position so that the baseline trace is at the bottom of the screen.

Now connect a DC power supply to channel 1 of the oscilloscope, using a BNC to standard lead adapter and two standard leads. Switch on the power supply and adjust the voltage output. When the power supply's voltage increases, notice that the trace on the oscilloscope rises. If the voltage is too high, the trace will rise above the top of the screen and disappear. At this point you will need to either lower the voltage or increase the setting for the number of volts/div.

With this knowledge, set the power supply to about 5 volts, and, with the voltage scale set at *1 volt/div*, record how many divisions (to the nearest tenth of a division) the baseline trace rises above the zero level. Also record the voltage scale of the scope. Connect a multimeter to the power supply and record the reading from the multimeter as well.

Next increase the voltage by 5 volt increments until you reach a voltage of 25 volts and record the number of divisions the baseline is above zero, the voltage scale setting, and the voltage reading from the multimeter for each. Feel free to change the oscilloscope's voltage scale between readings as needed.

P3. Repeat P2, but this time starting with a reading of 0.5 volts on the power supply and the voltage scale set at *.1 volts/div*. Increase the voltage in 0.5 volt increments until you reach 2.5 volts. Again feel free to change the oscilloscope's voltage scale as needed.

ATTENTION!

When using the oscilloscope, always make sure to record the scale setting for both the horizontal (time) scale and the vertical (voltage) scale for <u>EVERY</u> reading.

NOTE!

In general, signals that exceed about eight times the VOLTS/DIV switch setting should not be applied to the scope, so either the expected signal must be anticipated before setting the switch, or the switch should initially be set at its maximum value and then reduced until the unknown signal trace is located.

P4. Next connect another DC power supply to channel 2 of the oscilloscope, but do not remove the first power supply from channel 1. Connect another multimeter to the second power supply. Change the VERTICAL MODE switch from channel 1 to channel 2 and use the vertical position switch to adjust the position of the new baseline trace to the bottom of the screen. Turn the second power supply on and adjust the voltage up and down and notice that the movement of the baseline on the screen is the same as it was for the connection to channel 1. (You do not need to record anything for this part.)

Now set the VERTICAL MODE switch to BOTH and locate and set the ADD-ALT-CHOP switch to ALT. The screen should display a baseline trace that flashes back and forth between that of channel 1 and that of channel 2. If there is only one trace, adjust the voltage scales of either channel or the voltages of either power supply until you see two traces. (If the flashing begins to cause eye strain, set the switch to CHOP to display both traces simultaneously.) Once two baseline traces are clearly visible, set the ADD-ALT-CHOP switch to ADD. A new trace will appear that is the sum of the channel 1 and channel 2 inputs.

Notice that one or both voltage scale knobs have an INVERT switch next to them. Pushing this switch will display the negative of that channel's input. If only one channel is set to INVERT, the ADD function will display the difference between the two voltage inputs.

Now set the switch to CHOP, set the two voltage scale knobs to the same setting, and set the two power supplies to the same voltage. Invert one of the channels and set the ADD-ALT-CHOP switch to ADD. Where is the new trace on the screen? Take note of its position. Set the switch to CHOP and change the voltage scale of only one channel. There should be two traces now. Set the switch to ADD. Where is the trace in comparison to the previous ADD trace? Is this behavior expected? Explain.

AC VOLTAGE MEASUREMENTS

P5. Disconnect the DC power supplies and set the oscilloscope to display input on channel 1. Set the AC-GND-DC switch for channel 1 to AC. Use the vertical position knob to position the trace along the line at the center of the screen. This will be the new zero level. Use a coaxial cable with BNC connectors to connect channel 1 of the oscilloscope to the output of a function generator. The output will be labeled HI or OUTPUT. On the function generator, set the FUNCTION switch to SINE WAVE and adjust the amplitude control to its maximum. Set the function generator's frequency to about 100 Hz.

You will notice that the oscilloscope has a third large knob similar to the knobs for voltage scale that has settings for the time scale instead. This knob sets the amount of time represented by each division on the horizontal axis of the oscilloscope screen. The times will be displayed in seconds (s), milliseconds (ms), or microseconds (µs). When using this scale, be careful to note the specific units as well as the number on the scale.

NOTE!

You may see an irregular looking RF signal on the oscilloscope if the wires are connected to the scope and not connected to the signal generator, but this should disappear when you connect to the generator.

NOTE!

Channel 2 of the scope could just as well have been used for these measurements if the VERTICAL MODE switch had been set to CH 2 and the power supply connected to the CH 2 input terminal.

For this part of the experiment, set the time scale to *2 ms/div*. Adjust the voltage scale of channel 1 until a sine waveform appears and covers about four vertical divisions from its peak to its trough. From the center of the waveform to its peak, count the number of **vertical** divisions it covers and record the voltage scale of channel 1. This will be used to find the peak voltage reading on the oscilloscope. Now disconnect the function generator from the oscilloscope and connect a multimeter to the function generator. Make sure the generator's frequency is unchanged and record the multimeter's voltage reading. This is the RMS voltage of the waveform.

AC FREQUENCY MEASUREMENTS

P6. Reconnect the function generator to channel 1 of the oscilloscope. Adjust the frequency of the function generator until two or three cycles of the sine wave are visible on the screen. Use the horizontal position knob to adjust the waveform so that the number of divisions covered by **one cycle** of the wave can be easily measured. (One cycle is the **horizontal** distance between two peaks or two troughs.) Record the number of divisions in one cycle, the setting of the time scale knob, and the frequency reading from the function generator.

P7. **Without changing the frequency of the function generator,** adjust the time scale of the oscilloscope to a different setting and record the setting and the number of divisions in one cycle. Make sure the frequency output of the function generator remains constant.

P8. Keeping the frequency on the function generator constant, try various time scale settings and notice what happens to the display on the oscilloscope as the time per division increases and decreases.

Remember that when using the oscilloscope for any measurement, the scale settings for both the time scale and the voltage scale must always be recorded. Without this information, your measurements are meaningless.

BEATS

P9. Obtain the baseline trace (P1). Using the BNC connector on the HI or OUTPUT terminal of the function generator, connect the output from the function generator to the CH 1 input terminal of the scope and, with a frequency setting of about 1 kHz on the function generator and the appropriate setting of SEC/DIV, obtain a stabilized sinusoidal display of about four divisions in amplitude on the scope. For convenience later, adjust the frequency so that the period corresponds to an integer number of grid divisions (say three) on the screen; a *0.2 ms/div* setting on the scope should be about right.

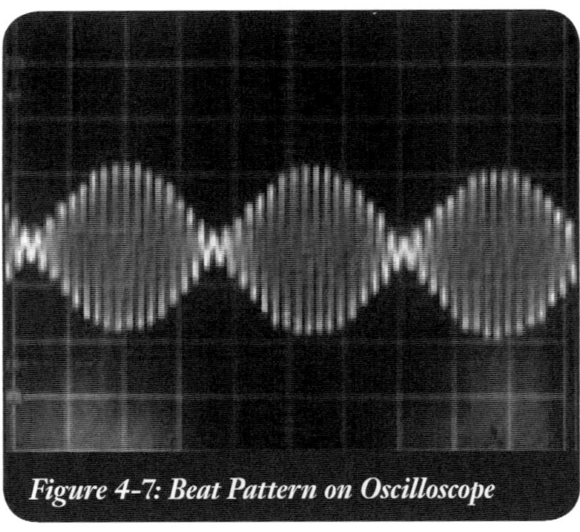

Figure 4-7: Beat Pattern on Oscilloscope

P10. Record the number of divisions required for one cycle and the setting on the SEC/DIV knob.

P11. Next, using a second BNC cable, connect the HI or OUTPUT terminal of a second function generator to the CH 2 input and obtain another stabilized signal of about the same magnitude and frequency as the first. As you change the VERTICAL MODE switch from CH 1 to CH 2 you should see approximately the same signal on either channel.

P12. Now set the VERTICAL MODE switch to BOTH so that the two signals are added together. Reset the SEC/DIV to about *2 ms/div*, change the frequency on the generator connected to CH 2 slightly until a clearly-defined beat signal is seen (see Figure 4-7). Use the screen grid and record the number of divisions for one beat cycle along with the reading from the SEC/DIV knob on the scope.

P13. Switch back to CH 2, adjust the time scale to an appropriate setting (**without** changing the frequency on the function generator), and record the number of divisions for one cycle and the reading from the SEC/DIV switch for that signal.

LISSAJOUS FIGURES

The SEC/DIV dial on the scope has a setting you have not yet explored: *xy*. This provides a display in which the vertical motion of the electron beam is controlled by the CH 1 input, as usual, but the horizontal motion is controlled by the CH 2 input rather than by the internal sweep mechanism.

A Lissajous[8] figure results from the superposition of two oscillatory motions occurring simultaneously at right angles to one another. Such a pattern should then result if CH 1 is used to display one sinusoidal wave form vertically while CH 2 is used to display another horizontally. The patterns can become very complex if the ratio of the two frequencies is not a rational number (ratio of integers). If, however, the ratio is rational, the resulting figures can provide a useful method for comparing frequencies of waveforms. For exam ple, if both oscillations have the same frequency (f) and phase but different amplitudes (A and B), the vertical (y) and horizontal (x) traces can be represented mathematically as

$$y = A \sin(2\pi f t) \tag{1}$$

and

$$x = B \sin(2\pi f t) \tag{2}$$

Eliminating t between the equations above gives

$$y = \left(\frac{A}{B}\right) x \tag{3}$$

This is the equation of a straight line with a slope of A/B.

The above situation occurs only if the vertical signal and the horizontal signal have the same frequency and the same phase. This is generally true only when the two signals both come from the same source. We will start with this simple case, then progress to somewhat more complicated combinations.

P14. Connect one function generator simultaneously to both CH 1 and CH 2 of the scope, by attaching the BNC t-connector (Figure 4-4) to the HI or OUTPUT terminal of the function generator, and attaching two BNC cords to the t-connector. Connect the other end of one BNC cord to CH 1 and the other to CH 2.

Set the frequency on the function generator to about 10 Hz, and set the SEC/DIV switch on the scope to *xy*. You should see a straight line on the screen. Select two points near the ends of this line and record the number of vertical divisions and the number of horizontal divisions between these two points. Also record the settings on the VOLTS/DIV knob for **BOTH** channels. (This information will be used to determine the slope of the line.)

P15. Move the MODE switch first to CH 1 and then to CH 2 and record the amplitude for the signal on each channel (be sure to record the number of divisions and the setting on the VOLTS/DIV knob for each channel).

Equal frequency signals in general will not give the simple straight line just observed, but rather will produce an elliptical or circular trace. For example, if the two signals have the same amplitude but differ in phase by $\pi/2$ radians, then

$$y = A \sin(2\pi f t) \tag{4}$$

and

$$x = A \sin\left(2\pi f t + \frac{\pi}{2}\right) \tag{5}$$

Note

[8] The French Physicist Jules Antoine Lissajous published his prediction and analysis of such superpositions in 1857.

By doing some algebra to eliminate t in Equations (4) and (5), you should be able to show that

$$x^2 + y^2 = A^2 \qquad (6)$$

Hint: Use the following identities from trigonometry:

$$\sin(a + b) = \sin a \cos b + \cos a \sin b \qquad (7)$$

$$\sin^2 a + \cos^2 a = 1 \qquad (8)$$

Equation (6) is the equation of a circle of radius A. Unfortunately slow drifts in phase and frequency prevent such Lissajous figures from being very stable; a circle will slowly rotate or evolve into a straight line or an ellipse and back to a circle.

P16. Now remove the t-connector and connect the HI or OUTPUT from one signal generator to CH1 and the HI or OUTPUT of **another** signal generator to CH 2. Make sure both generators are set to produce sine wave patterns and adjust both frequencies to about the same value. When the frequencies are identical you should see a very slowly rotating circle as described above. Next reduce the frequency of one generator until it is half the frequency of the other; sketch the resulting pattern. Continue varying the generator frequencies until recognizable patterns are obtained and record the corresponding frequencies. See if you can obtain patterns for frequency ratios of 3:1, 4:1, and 2:3.

CALCULATIONS

DC VOLTAGE MEASUREMENTS

C1. For each voltage setting in P2 and P3, multiply the number of divisions above the zero level by the voltage scale to obtain the oscilloscope's reading of the voltage. Compare the voltage readings from the oscilloscope and the multimeter by calculating the percent difference between the two values. Do these readings agree (taking experimental uncertainty into account)?

AC VOLTAGE MEASUREMENTS

C2. Multiply the number of divisions above zero level by the voltage scale to obtain the oscilloscope's peak voltage reading from P5, Since the multimeter's reading is the RMS voltage, it must be converted to peak voltage using the equation: $V_{peak} = \sqrt{2}\, V_{RMS}$. Once both peak voltages have been obtained, compare them by calculating their percent difference. Do they agree (again taking experimental uncertainty into account)?

AC FREQUENCY MEASUREMENTS

C3. Multiply the number of divisions in one cycle for the waveform in P6 by the time scale to obtain the period (T) reading of the oscilloscope. Convert the period to frequency using the equation $f = 1/T$. Compare the frequency values of the oscilloscope and function generator by calculating percent difference. Do the values agree (taking experimental uncertainty into account)?

C4. Calculate the period from the oscilloscope reading in P7 and use it to calculate the frequency. Compare this frequency reading to the frequency reading from C3. How did the frequency readings vary as the time scale was changed? Is this the expected result if only the time scale changed between the readings?

BEATS

C5. Use the method of C3 to determine the frequency for the CH 1 signal, the CH 2 signal and for the beats that were obtained by adding those two together in P12. According to theory, the beat frequency should be equal to the difference between the two frequencies that add to produce the beats. Calculate the expected frequency from the values obtained for the CH 1 and CH 2 frequencies (P10 and P13) and compare that value to the one obtained from the scope reading of the beat pattern.

LISSAJOUS FIGURES

C6. Using the data from P14 for the number of vertical and horizontal divisions between the two points you selected on the straight line you obtained by feeding the same signal into both channels, calculate the slope of the line. This is the value of A/B in Equation (3). Examination of Equations (1) and (2) shows that A is the amplitude of the signal on CH 1 and B is the amplitude on CH 2. Use the values obtained from the separate displays (P15) to determine another value for the slope of the line produced when the two signals are combined. How do the two values compare?

C7. Do the calculations to obtain Equation (6) by eliminating t from Equations (4) and (5). As an additional exercise to test your understanding of Lissajous figures and superposition, describe the patterns to be expected if the two oscillations producing the figures have equal frequencies and amplitudes but differ in phase by 0, $\pi/2$, π, and $3\pi/2$ radians. Be sure to give the specific form of the function (e.g., straight line, parabola, circle, ellipse) and give information pertinent to that particular form (e.g., slope, radius, length of semi-major axis).

$$R = \frac{\rho \ell}{A}$$

R = resistance (Ω)

ρ = resisitivy $(\Omega \cdot m)$

ℓ = length

A = cross-sectional area

ρ Copper $= 1.68 \times 10^{-8}\ \Omega \cdot m$

Ohm's Law: $V = IR$

.4 Ω, gator clamps

EXPERIMENT 5

Circuits and Circuit Elements

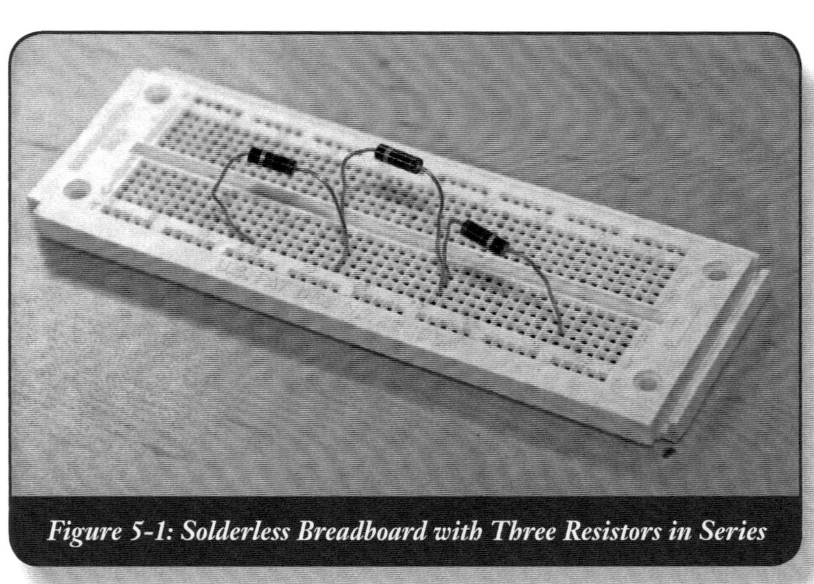

Figure 5-1: Solderless Breadboard with Three Resistors in Series

INTRODUCTION

In this experiment properties of resistors and capacitors will be examined, and some simple DC circuits containing these elements will be constructed and analyzed using the multimeter and the oscilloscope.

Using the multimeter, the dependence of resistance on length and cross-sectional area of copper and nickel-silver will be determined. Then the resistance of a commercial resistor will be determined by measuring the current as a function of voltage, using the multimeter to determine the current and the oscilloscope to measure the voltage. A non-linear (i.e., non-ohmic) device, a common lightbulb, will be investigated in the same way. Commercial resistors will then be assembled in series and parallel combinations on a solderless breadboard, and the multimeter used to verify the predictions of theory.

For capacitors, the multimeter will be used to measure the value of several commercial examples, and again the breadboard used to investigate series and parallel combinations.

PROCEDURE

P1. Recall that to use the multimeter to determine resistance the red test lead should be connected to the **V-Ω-Hz** jack and the black test lead to the **COM** jack. Use the meter to measure the resistance of each coil of wire on the mounted set; record the corresponding length, diameter, and type of metal for each coil. The diameters marked on the mounting base are only approximate; coils labeled *0.025* actually have diameters of 0.02535 inches and those labeled *0.012* have diameters of 0.01264 inches.[9] Note that the diameters are given in British units while the lengths are given in SI units.

P2. Commercial resistors of the type used in common circuits come in a wide variety of construction, packaging, and power tolerances. Their nominal resistance values are usually indicated by a set of colored bands encircling the resistor; sometimes there are four bands, sometimes five. A chart illustrating this color code can be found in Experiment 1 of this manual. From the assortment of resistors you have been given, select one with a nominal value in the range of 100 to 200 ohms. Set up the circuit shown in Figure 5-2 (b), using the

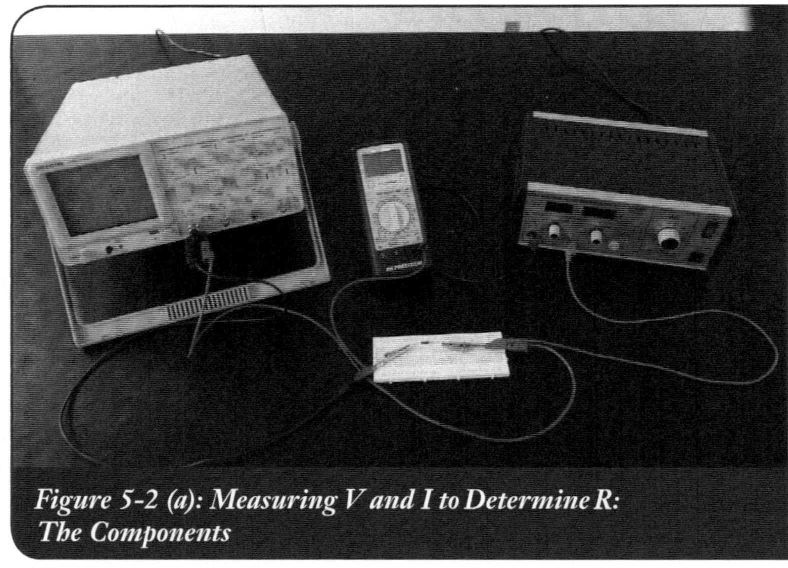

Figure 5-2 (a): Measuring V and I to Determine R: The Components

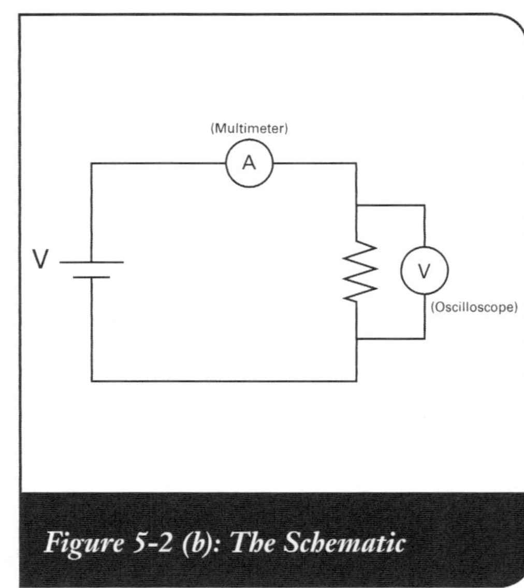

Figure 5-2 (b): The Schematic

Note
[9] Brown & Sharpe wire gauges 22 and 28, respectively.

multimeter as the ammeter and the oscilloscope as the voltmeter. First set up the loop that contains the power supply, the ammeter (multimeter) and the resistor. Then connect the voltmeter (oscilloscope) across the resistor. Ask your instructor if you need help to follow the circuit diagram. Use the **mA-μA** terminal on the multimeter for the positive lead and the **COM** terminal (as always) for the negative lead. Set the current range to *40 mA*. Set the controls of the oscilloscope to obtain a baseline at the bottom of the screen on the oscilloscope (with the power supply off). Set the **VOLTS/DIV** initially at 0.1. (You will need to change the scale as you increase the voltage.)

With the **COARSE** and **FINE** controls of the power supply fully counterclockwise (*off*), turn on the supply. Using the **COARSE** and **FINE** controls, slowly bring the voltage up to 0.1 volts, as determined from the scope. Read and record the current from the multimeter. Continue increasing the voltage, reading the corresponding current at intervals of 0.1 volts until 1.0 volt is reached, then increase the interval to 0.5 volts and continue readings until the current at 25 volts is obtained. Be sure to read the voltage from the oscilloscope, **not** the power supply. Change the scale on the oscilloscope as needed as the voltage gets larger, being careful to note the scale used for each measurement.

P3. Replace the resistor used in P2 with the small Christmas tree bulb. (Alligator clips can be attached to the two small wires that stick out from the bulb.) Repeat the procedure of P2, but using potentials ranging from 0.1 volt to 2.0 volts, in 0.1 volt intervals. **<u>DO NOT</u>** exceed 2.0 volts in this part.

P4. Begin by closely examining the breadboard. Such boards permit quick temporary connections to be made between circuit elements, and provide a convenient means of testing various configurations before a design is finalized. The top surface of the board is perforated with holes into which the leads from resistors, capacitors, etc., can be inserted. Some of these holes are interconnected beneath the board. Figure 5-3 shows which holes on the breadboard are connected to one another. (Everything along a continuous line is connected.)

NOTE!

Some of the breadboards are not labeled explicitly but the connections are the same as for the boards that are labeled.

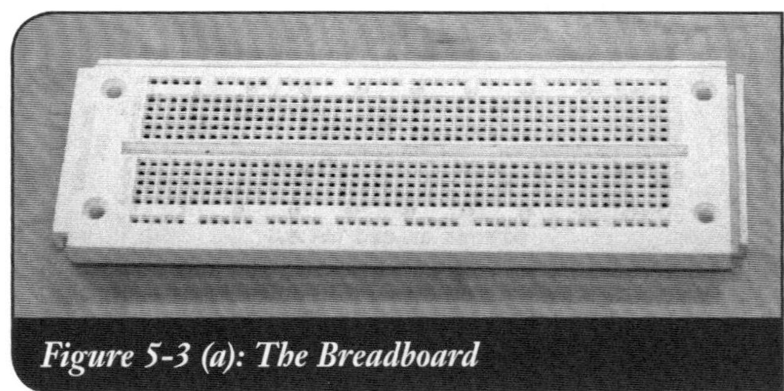

Figure 5-3 (a): The Breadboard

Figure 5-3 (b): Connections on a Breadboard

Select two resistors and connect them in series on the breadboard, but do not connect a power supply to the combination (see Figure 5-4). Use the multimeter to measure the resistance of the combination. Also measure and record the resistance of each individual resistor. Then rearrange the same two resistors in a parallel combination and again use the multimeter to measure the resistance of the combination. (See Figure 5-5.) Record the values of the individual resistances and of the series and parallel combinations. Be sure to use the measured values of the resistors, not the values obtained from the color code.

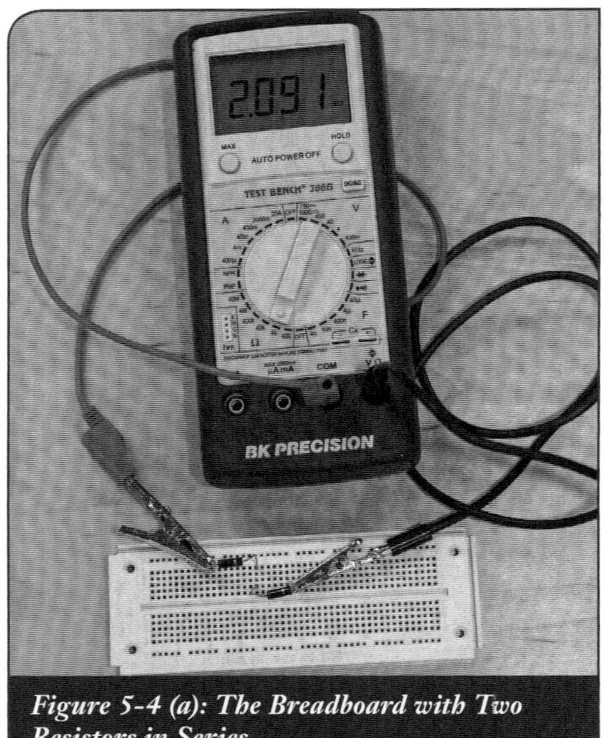

Figure 5-4 (a): The Breadboard with Two Resistors in Series

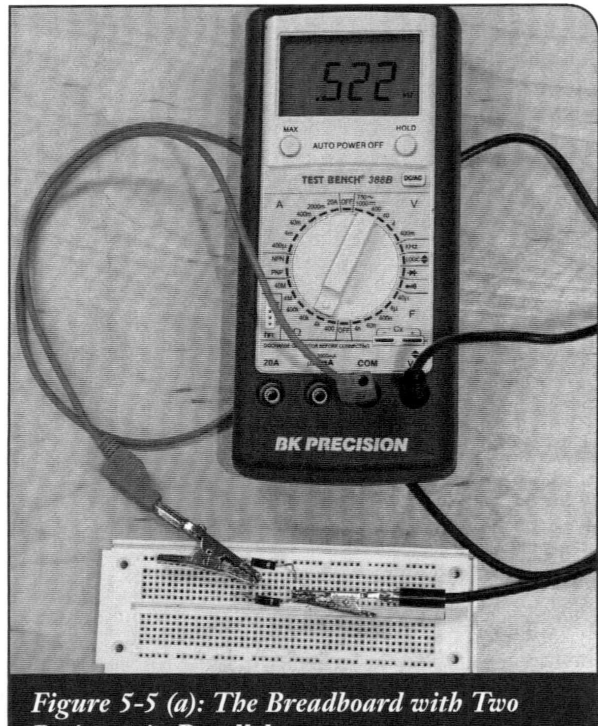

Figure 5-5 (a): The Breadboard with Two Resistors in Parallel

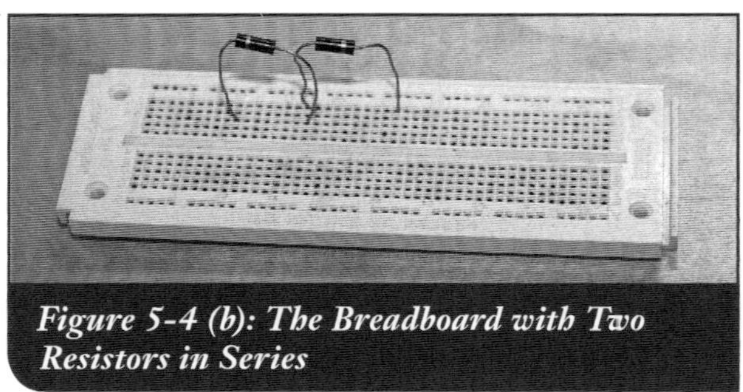

Figure 5-4 (b): The Breadboard with Two Resistors in Series

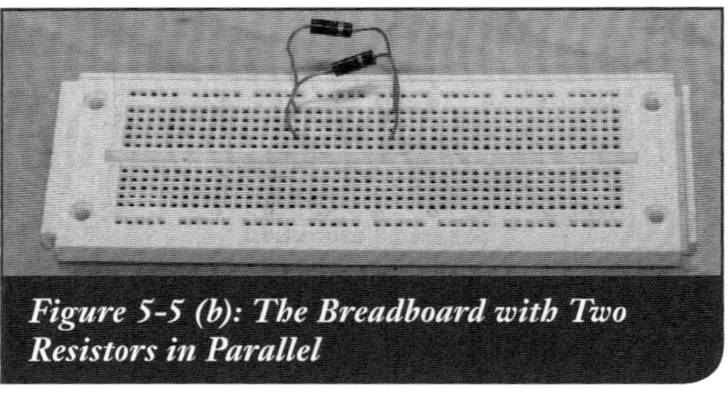

Figure 5-5 (b): The Breadboard with Two Resistors in Parallel

P5. Capacitors, like resistors, come in many types of packages. (See Figure 5-6.) In some a color code is used to indicate the value, but in others the value is labeled directly on the body of the capacitor.

To measure capacitance with the multimeter the test leads are not used. Instead the leads of the capacitor are inserted directly into the slotted **Cx** test jacks. (See Figure 5-7.) It is important that external voltages never be connected to these jacks, so before capacitor leads are inserted they should be touched together to insure that there is no residual charge on the capacitor. Select one of the capacitors you have been given, briefly touch its leads together, then insert them into the Cx jacks and, with the range selector set to the appropriate capacitance range, record the reading.

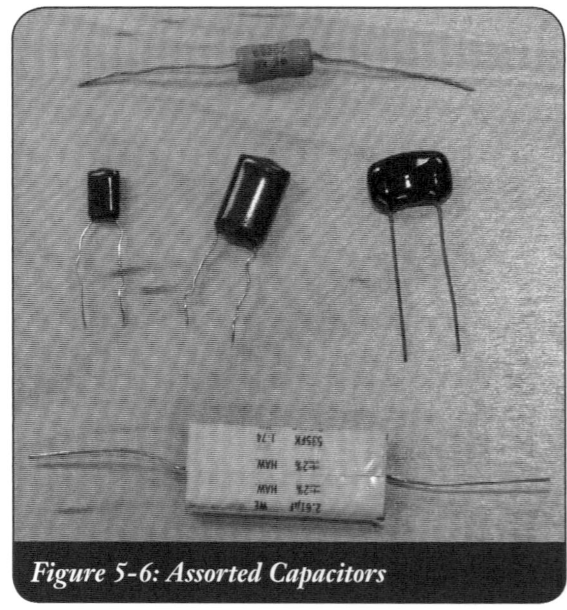

Figure 5-6: Assorted Capacitors

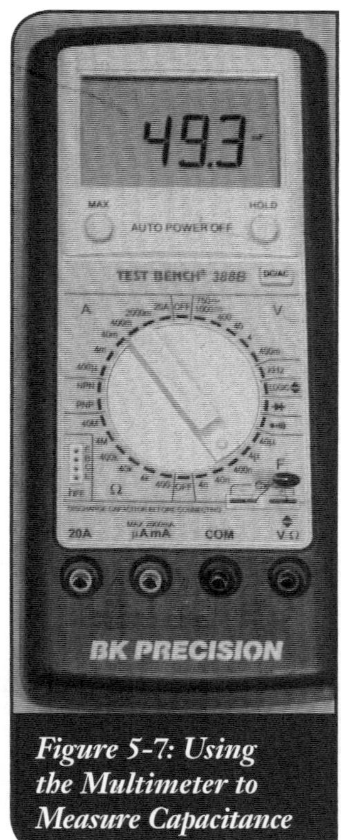

Figure 5-7: Using the Multimeter to Measure Capacitance

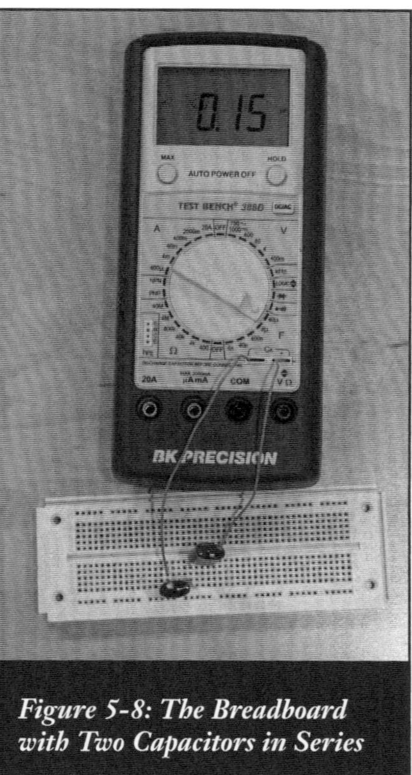

Figure 5-8: The Breadboard with Two Capacitors in Series

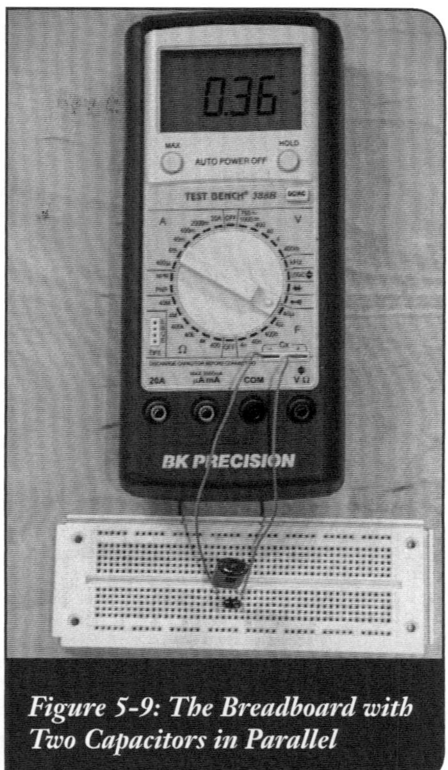

Figure 5-9: The Breadboard with Two Capacitors in Parallel

Next, select two capacitors and, using the breadboard, assemble them in series (Figure 5-8). A kit containing wires of various lengths is provided to facilitate connections. Since leads from the capacitors must be inserted directly into the meter test jacks, use jumper wires from the wire kit to connect the series combination to the meter **Cx** jacks, and record the result. Don't forget to measure and record the individual values as well. Repeat the procedure with the capacitors connected in parallel (Figure 5-9).

CALCULATIONS

C1. Note that of the resistance spools used in P1, spools 1 and 3 are both copper wire and have the same diameter, but spool 3 is twice as long as spool 1. Compare the ratio of the measured resistances of these two spools with the ratio of their lengths. Make a similar comparison between spools 2 and 4. How do these results compare with the predictions of theory ($R = \rho L/A$)?

Spools 1 and 2 differ only in diameter. The same is true of spools 3 and 4. Compare the ratio of their measured resistances with the inverse ratio of the squares of their diameters, and compare your results with the predictions of theory.

Using the measured values of R for each spool, calculate the resistivity for each spool. Average the results for spools 1 through 4 to obtain the resistivity of copper. Compare your results for both copper and the nickel-silver alloy with those reported in the *Handbook of Chemistry and Physics*. Can you account for any differences?

C2. From the data obtained in P2, make a graph of V vs. I. Use the method of least squares to obtain the slope and its uncertainty, and compare your result with the nominal value of the resistance and its uncertainty as determined from the color bands.

C3. From the data of P3, make a graph of V vs. I for the light bulb. Is the graph linear? If not, over what range of values does the resistance vary?

C4. For the parallel and series combinations of resistors used in P4, compare the measured resistance of the combination with the value calculated from the measured values of each component. (For two resistors R_1 and R_2 in series the expected series value R is given by $R = R_1 + R_2$; the expected parallel value is given by $1/R = 1/R_1 + 1/R_2$.)

C5. For the series and parallel combinations of capacitors used in P5, compare the measured capacitance of the combination with the value calculated from the measured values of each component. (For two capacitors C_1 and C_2 in series, the expected series value is given by $1/C = 1/C_1 + 1/C_2$; the parallel value is given by $C = C_1 + C_2$.)

EXPERIMENT 6

Resistor and Capacitor Networks

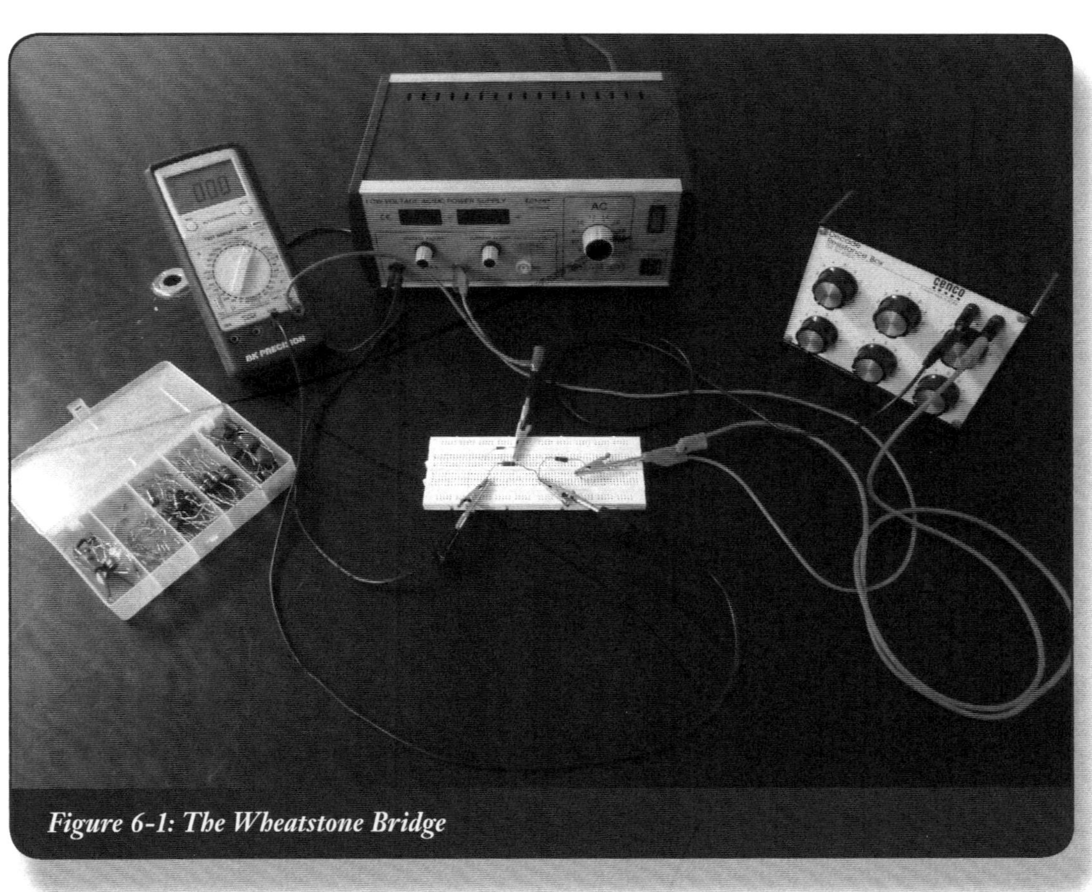

Figure 6-1: The Wheatstone Bridge

INTRODUCTION

The previous experiment consisted of an investigation of various circuit elements and their combination to construct simple circuits. Those concepts will now be used to develop more complex combinations.

Two resistor networks and one capacitor network will be assembled on the breadboard. As before the multimeter will be used to verify the predictions of theory for these networks.

Two circuits of special interest, the Wheatstone bridge and the RC-circuit will then be assembled and their behavior will be investigated using an oscilloscope.

PROCEDURE

P1. Choosing convenient values of resistances, assemble the combination shown in Figure 6-2 and measure the equivalent resistance. (Use resistors that are all of the same order of magnitude; i.e., the third stripe indicating the power of ten is the same for each. Also measure and record the values of each component resistor. Include the diagram in your report, with the value of each resistor labeled. Be sure to use the values of resistance <u>measured</u> by the multimeter. Do not rely on the color code on the resistors.

P2. Using the breadboard, assemble the resistor network shown in Figure 6-3. Use resistors of all equal value. Again include the diagram and measure and record the resistance of each of the individual resistors and of the combination.

P3. Next, assemble and measure the individual capacitances and the equivalent capacitance for the combination shown in Figure 6-4. As before include a diagram in your report and label the individual capacitances on the diagram. (Don't forget that you must use the slots rather than the terminals on the multimeter to measure capacitance.)

P4. Next consider the Wheatstone bridge (Figure 6-5). The detector must be a device which responds to a potential difference or flow of current; it is said to bridge the upper and lower resistance branches. If the detector senses no potential difference between points A and B, then it is easily verified that

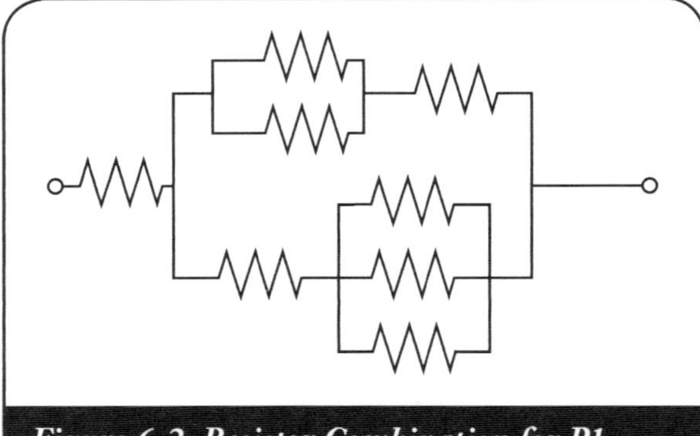

Figure 6-2: Resistor Combination for P1

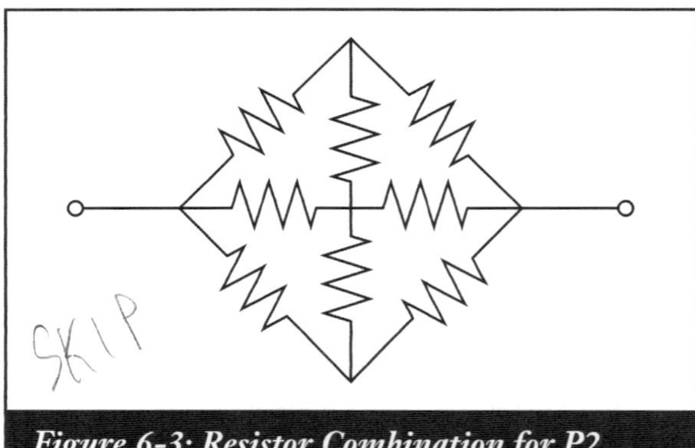

Figure 6-3: Resistor Combination for P2

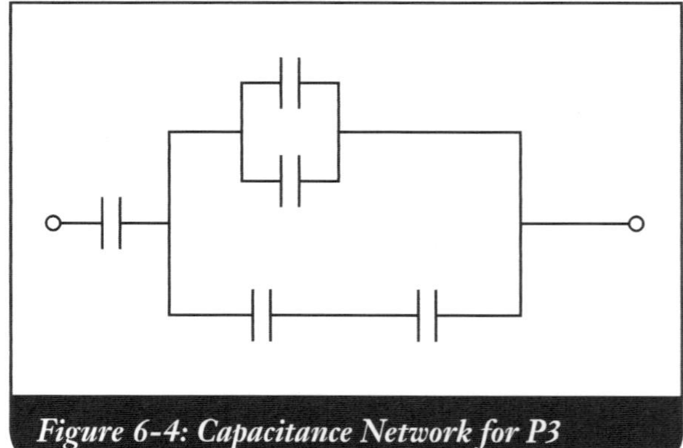

Figure 6-4: Capacitance Network for P3

$$\frac{R_a}{R_c} = \frac{R_b}{R_d}$$

and the bridge is said to be balanced. The principal application of this circuit is based on using three known resistances to determine a fourth (e.g., if R_a, R_c, and R_d are known, then $R_b = R_a(R_d/R_c)$ when the bridge is balanced.

Using the breadboard, assemble the bridge. For R_a, R_c, and R_d use commercial resistors; measure and record their values with the multimeter. Convenient values for R_a and R_c would be between 100 and 500 ohms, and for R_d between 1000 and 2000 ohms. Use the decade resistance box for R_b; connect it to the breadboard using banana cords and alligator clips. R_b will be adjusted to bring the bridge into balance, and the resulting value compared with the prediction of the balance equation. Again using banana cords and alligator clips, connect the bridge to the DC Power Supply. Make sure that the **COARSE** and **FINE** controls are fully counterclockwise and that the supply is turned off. For the detector, use the multimeter with the scale set on

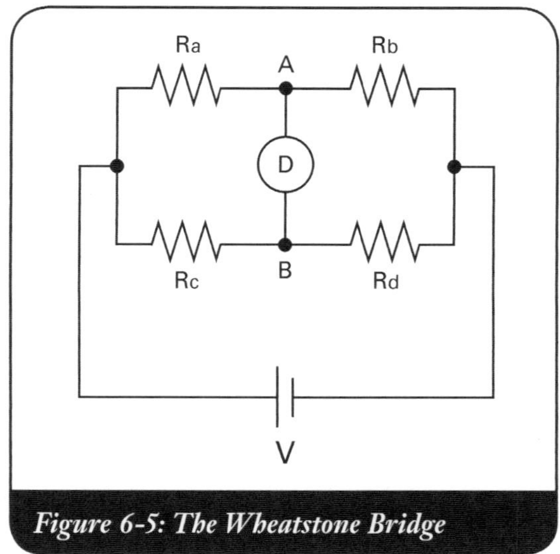

Figure 6-5: The Wheatstone Bridge

40 V; the meter probes can be connected to the breadboard (points A and B in Figure 6-5) with alligator clips, or the probes may simply be held in place. As usual, the red probe connects to the **V-Ω-Hz** jack and the black probe to the **COM** jack.

Set the Decade Resistance Box to about 3000 ohm. Now **turn on the power supply** and gradually increase the voltage to about 10 volts. Note the voltage reading on the multimeter, then decrease the decade resistance until the meter indicates no potential difference between points A and B. Record the decade resistance.

Next, the multimeter will be replaced with the oscilloscope and the experiment repeated in order to compare the sensitivities of the two methods of detection.

Begin by removing the multimeter connections to points A and B and replacing them with connections to channel 1 on the oscilloscope. Set the **AC-GND-DC** switch to **GND** and adjust the position so that the trace is in the middle of the screen. Then return the switch to the **DC** position. Begin your measurements with the decade resistance box again at 3000 ohms, and set the **VOLTS/DIV** on the scope to 5 or 10 volts. Note the initial voltage reading from the scope, then gradually decrease the decade resistance until the scope trace returns to ground level; at this point reduce the **VOLTS/DIV** reading to increase the scope sensitivity, and continue reducing the decade resistance. Continue this process until you feel certain, based on the reading of the scope, that the potential difference between points A and B is truly zero. Record the setting of the decade resistance.

P5. As a last circuit example, consider the series RC circuit (Figure 6-6).

When switch S_1 is closed, current begins to flow through R with no opposition from the capacitor because no charge yet resides there. But as charge accumulates on the capacitor a potential difference develops across the capacitor plates in opposition to the applied voltage. As this charge accumulates it eventually brings the current to a halt. The time required for the current to become zero is actually

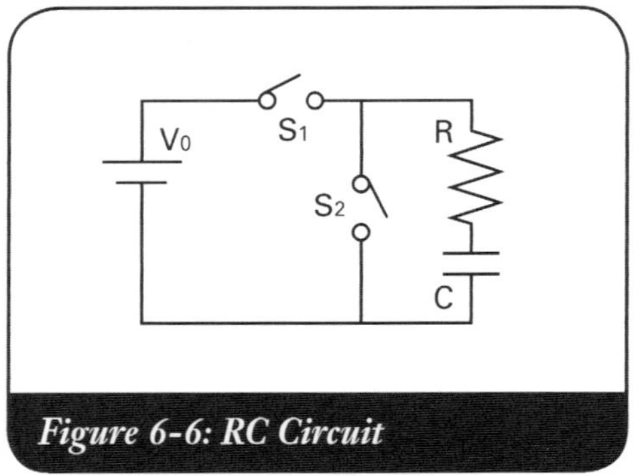

Figure 6-6: RC Circuit

infinite, but the current approaches zero exponentially in time with a characteristic time constant RC, the time required for the current to fall 63% from its initial value. If after the current has become essentially zero, switch S1 is opened and S2 closed, the capacitor discharges through the resistor again with a time dependence characterized by the same time constant RC. Since the potential across the resistor is proportional to the current through the resistor, the time dependence of this potential is also characterized by RC (Figure 6-7). When the capacitor is charging it is easily shown, by applying Kirchhoff's loop theorem, that

$$V_R = V_0 e^{-t/RC} \text{ and } V_C = V_0(1 - e^{-tRC})$$

and when discharging that

$$V_R = -V_0 e^{-t/RC} \text{ and } V_C = V_0 e^{-tRC}$$

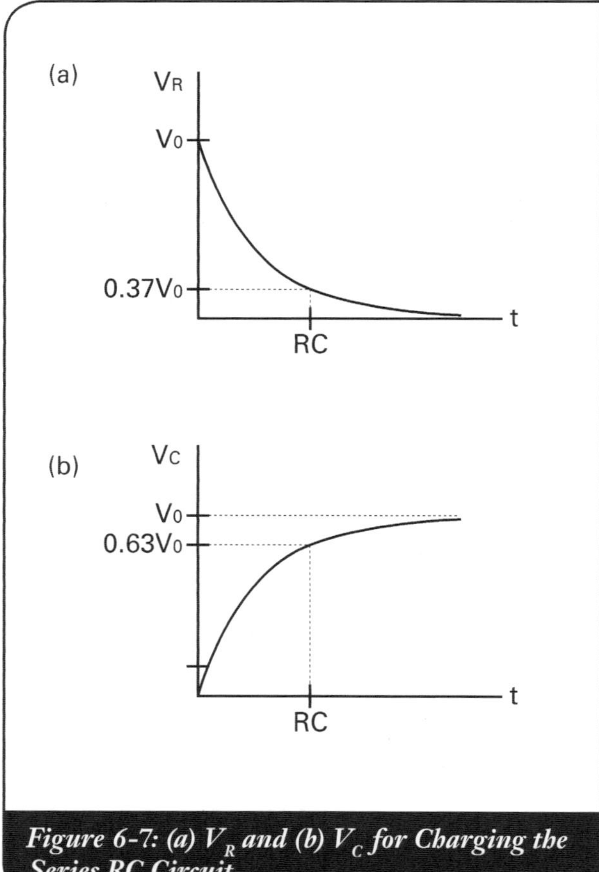

Figure 6-7: (a) V_R and (b) V_C for Charging the Series RC Circuit

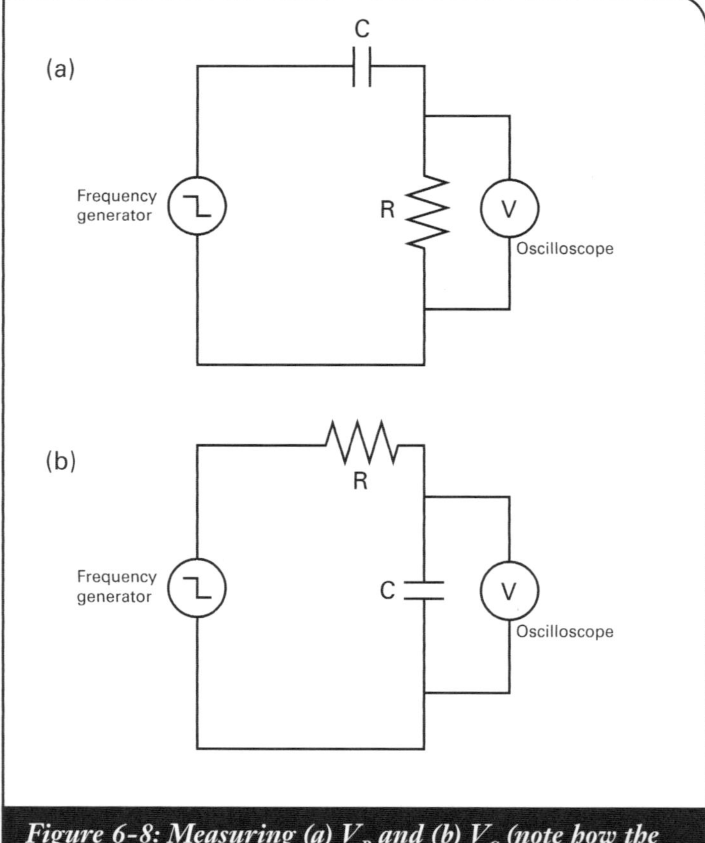

Figure 6-8: Measuring (a) V_R and (b) V_C (note how the connections are changed from (a) to (b))

You will assemble the circuit of Figure 6-6 using the breadboard and a typical commercial resistor and capacitor, but using a square wave from the frequency generator to accomplish the necessary switching (Figure 6-6). The square wave output oscillates between some amplitude +A and –A. When the amplitude is positive the RC circuit sees an applied voltage, equivalent to closing switch S_1 in Figure 6-6, and when the output goes negative the reverse effects occur. For this scheme to work properly, however, the period of the square wave must be large compared to the time constant of the RC circuit (why?). Begin by placing a resistor and capacitor in series on the breadboard. A reasonable value for R would be around 1000 ohms and for C around 50 nF, which would give a time constant of 50 ms, but select any combination you wish; remember that you want to be able to view at least one full charging cycle on the oscilloscope.

Next, using a banana-to-BNC adaptor at the 2V output terminal of the signal generator, connect the generator to CH 1 of the scope and, by plugging one set of banana cords into another at the generator output, simultaneously connect the output to the RC circuit (Figure 6-9). Select a square wave output with a period large compared to the expected time constant (500 Hz would be a reasonable frequency to use for a circuit with a 50 ms time constant.) Now connect the oscilloscope probe to **CH 2** and use it to determine the potential across the resistor, as shown in Figure 6-8 (a). From the SEC/DIV setting on the scope, determine the time constant by counting the number of horizontal divisions on the screen required for the trace to fall 63% of its full height (or to rise within 37% of its height). Manipulating the position controls and adjusting the SEC/DIV setting will facilitate this measurement.

After determining the time constant, vary the period of the square wave and note the effect on the response of the circuit; choose periods both small and large compared to the time constant, and draw rough sketches to illustrate your observations. Use the **BOTH** setting of the **VERTICAL MODE** switch on the scope to display the square wave and the response simultaneously.

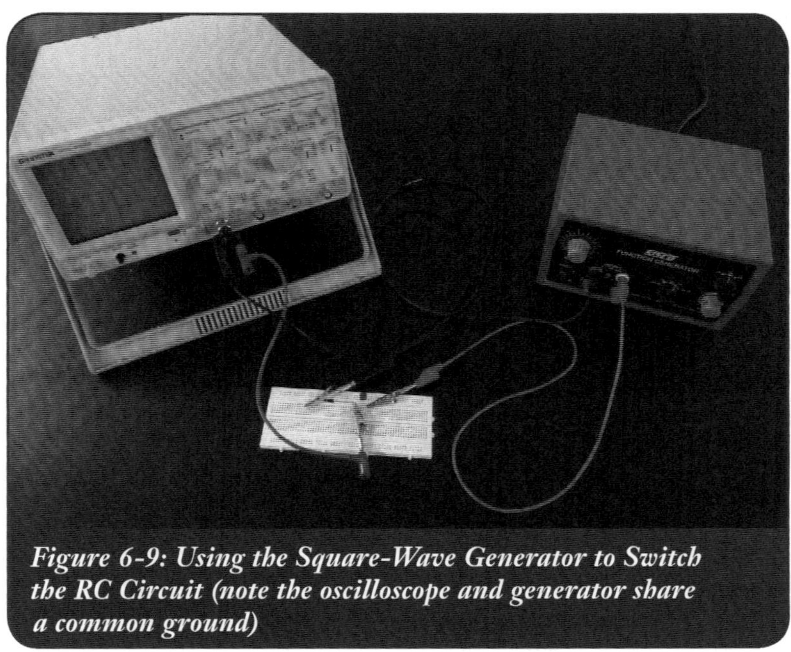

Figure 6-9: Using the Square-Wave Generator to Switch the RC Circuit (note the oscilloscope and generator share a common ground)

Next, reverse the positions of the generator leads on the resistor and capacitor, and use the probe to display the voltage across the capacitor (Figure 6-8 (b)). Again determine the time constant and observe the effect of changing the period of the square wave.

Finally, observe the effect on the display of changing the frequency of the signal generator, particularly as the generator frequency becomes large compared to the inverse of the time constant. Sketch the general character of the display for frequencies both large and small compared to the inverse of the time constant.

CALCULATIONS

C1. Compare the experimentally obtained value of the equivalent resistance for the resistor network in P1 with the value calculated from the theoretical equations for series and parallel resistor combinations.

C2. Repeat C1 for the resistor network in P2.

C3. Compare the experimentally obtained value of the equivalent capacitance for the capacitor network of P3 to the theoretically obtained value.

C4. From the data of P4 compare the values of the decade resistance when the bridge is balanced with the expected value calculated from $R_a(R_d/R_c)$. Recall that you made this determination first using the multimeter and then with the scope; which detector resulted in greater precision? Which yields a value closer to the theoretical value?

It should be pointed out that using either the multimeter or the oscilloscope as a detector in the Wheatstone bridge obscures one of the principal advantages of the circuit. The only requirement of the detector is that it provide a zero reading when there is no potential difference across the bridge, i.e., that it be capable of providing a *null* result. This important property eliminates any need for a *calibrated* detector, and a sensitive galvanometer would more likely be used in practice.

C5. *From* the measurements of P5, compare the value of the time constant t of the RC circuit as determined from the oscilloscope with that calculated from the individually measured values of R and C (t = RC). Discuss briefly your observations on the effect of varying the period of the square wave relative to the time constant.

Comment: *Using calculus it can be shown that when the period of the square wave is large compared to the time constant, the time dependence of the voltage across the resistor (Figure 6-7 (a)) is the derivative of the square wave; when the period of the square wave is small compared to the time constant, the time dependence of the voltage across the capacitor Figure 6-7 (b)) is the integral of the square wave. If you are familiar with the concepts of the derivative and integral, you might consider whether the observations made in P5 are consistent with these predictions.*

Notes

Purpose To determine the charge to mass ratio...

$$F = q\vec{v} \cdot \vec{B} = \frac{mv^2}{r}$$

$$B = \frac{8\mu_0 Ni}{\sqrt{125}\, a}$$

a = radius of circle

N = number of loop

$\rightarrow$ i = current in coils

$$\tfrac{1}{2} mv^2 = eV$$

e = electron charge

v = speed

$\rightarrow$ V = voltage

m = mass of electrons

$\rightarrow$ r = radius of circle of e^-'s

a = radius of helmholtz coil

1.) $\dfrac{mv^2}{R} = evB \cdot R$

$\dfrac{mv^2}{m} = \dfrac{(evB)R}{m}$

$\sqrt{v^2} = \dfrac{(evB)R}{m}$

$\tfrac{1}{2} m \left(\dfrac{eBR}{m} \right) = eV$

$V = eBR$

2.) $B = \dfrac{8\mu_0 Ni}{\sqrt{125}\, a}$

$B = \dfrac{8\left(4\pi \times 10^{-7}\ Tm/A\right)(130)i}{\sqrt{125}\,(.15)}$

$$B = \frac{1.306902544 \times 10^{-3}\ i}{1.667705}$$

$(B)(1.667705) = \left(1.306902544 \times 10^{-3}\right)i$

EXPERIMENT 7

Moving Charges in Magnetic Fields:
Determining the Charge to Mass Ratio for the Electron

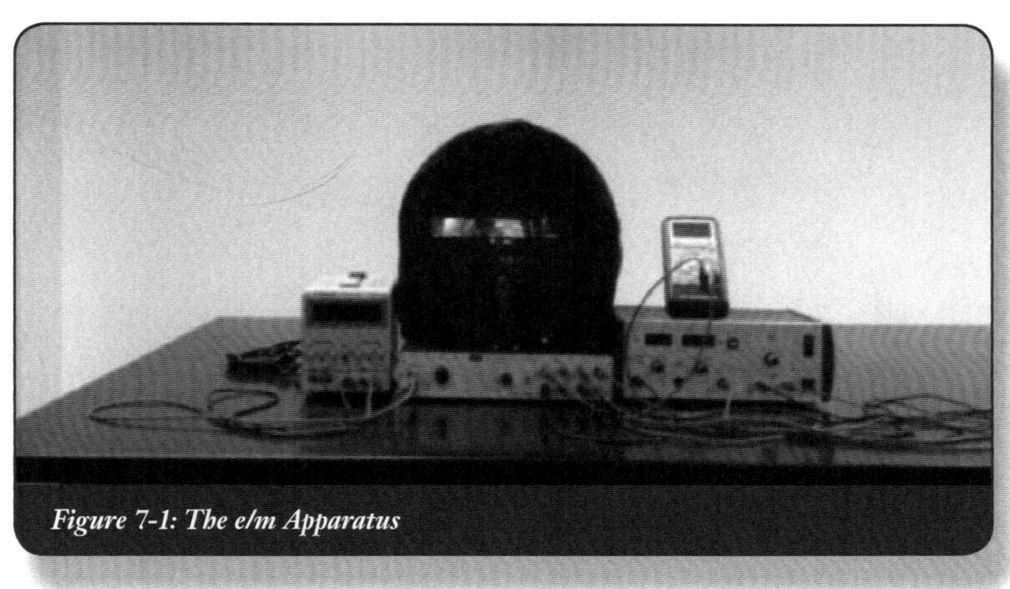

Figure 7-1: The e/m Apparatus

INTRODUCTION

The magnetic field **B** is defined in terms of the force experienced by a moving charge q:

$$\mathbf{F} = q\mathbf{v} \times \mathbf{B} \tag{1}$$

While this looks more complicated than the corresponding defining equation for the electric field in terms of the force experienced by a stationary charge, $\mathbf{F} = q\mathbf{E}$, the apparent complication is due only to the necessity of incorporating the velocity of the charge into the definition, and one definition is no more complicated conceptually than the other. The "cross-product" notation used in Equation (1) might be unfamiliar to you, but it simply means that the magnitude of the force is given by q v B sin θ, where θ is the angle between the velocity and field vectors, and that the direction of the force is given by the right-hand rule.

In this experiment a beam of moving electrons, made visible by the passage of the electrons through helium vapor in an enclosed glass tube, is deviated into a circular path by the application of a magnetic field. The radius of this path depends on, among other parameters, the ratio of the charge to the mass of the electrons, and hence measurement of the radius permits calculation of the ratio. The magnetic field is produced by passing current through two large coils, known as Helmholtz coils due to their particular geometry, and it is visually satisfying to see how the path of the electrons changes as the field is changed.

The Helmholtz coil geometry consists of two coils of wire, each of radius **a** (about 0.15 m in this apparatus), their planes parallel to one another and separated by a distance equal to the radius **a**. With current going the same direction in each coil the fields produced by each coil add at the center of the apparatus; it is not difficult to show that a current-carrying coil produces a field **B** at a point on the axis of the coil a distance **d** from the center which is given in magnitude by the expression

$$B(d) = \frac{\mu_0 i\, a^2}{2\sqrt{(a^2 + d^2)^3}} \tag{2}$$

and which points along the axis of the coil, either toward or away from the coil depending on the direction of the current. Multiplying this result by the number N of loops in each coil, using $d = a/2$, and multiplying the result by a factor of two to account for both coils, the field at the center of the apparatus is given by

$$B = \frac{8\,\mu_0 N\, i}{\sqrt{125}\ a} \tag{3}$$

where all quantities are in SI units. Because of the particular geometry of the Helmholtz coils, Equation (3) can also be used to find the field short distances along the axis on either side of the center point.[10] When a beam of electrons moving with velocity **v** perpendicular to **B** enters this field the electrons experience a force **F** given by Equation (1); since this force is perpendicular to **v** (and to **B**), the resultant motion is circular of radius **R** with an acceleration v^2/R. The electrons acquire the velocity **v** by being accelerated through a potential difference **V.** This is accomplished by producing the electrons from a straight filament surrounded by a coaxial cylindrical anode containing a narrow slit.

Note

[10] See, for example, Halliday & Resnick, *Fundamentals of Physics* (3rd ed.), Chapter 31, Exercise 65 and Problems 69 and 70, or Tipler, *Physics* (3rd ed.), Chapter 25, Problems 53 and 62, and photo, page 811.

The filament is heated by passage of a current through it; electrons are emitted and then accelerated by the potential difference **V** applied between the filament and the anode. In terms of this potential difference, the kinetic energy of the electrons as they enter the field **B** can be written as

$$\frac{m\,v^2}{2} = e\,V \qquad\qquad (4)$$

m being the electronic mass and **e** the charge. Thus the acceleration is **2eV/mR,** and Newton's second law, using Equation (1) for the force and (3) for the field, can be written

$$\frac{e}{m} = \frac{250\,V\,a^2}{64\,\mu_0^2\,N^2\,i^2\,R^2} \qquad\qquad (5)$$

N and **a** are constants characteristic of the Helmholtz coils (**N** = 130 and **a** is determined by measurement with a meter stick), **V** and **i** can be determined directly by measurement with appropriate meters, and **R,** the radius of the electron path, is determined by observing where the path crosses known positions marked within the electron tube.

PROCEDURE

P1. Flip the toggle switch on the front of the e/m apparatus to the e/m MEASURE position and turn the current adjust knob for the Helmholtz coils (on the left end of the front panel of the apparatus) to the OFF position.

P2. Set up the anode circuit by using electrical lead wires to connect the black DC terminals labeled 500 V on the large PASCO® power supply to the corresponding terminals on the front panel of the e/m apparatus. (The terminals for the anode circuit are the second set of terminals from the right on the front panel. See Figure 7-2.)

Connect a multimeter to measure the voltage on the anode, by connecting the VOLTS and COM terminals of the multimeter to the corresponding terminals of the leftmost of the four sets of terminals on the right side of the front panel of the apparatus. Make sure the multimeter is set on DC and set it to read up to 300 volts.

P3. Connect the yellow AC terminals of the same power supply to the heater, using the rightmost set of terminals on the front panel of the apparatus.

P4. Connect the white G^{W}INSTEK power supply and another multimeter to measure current in series with the Helmoltz coil. First connect the positive terminal of the power supply to the AMPS terminal of the multimeter. Then connect the COM terminal of the multimeter to the positive terminal on the left side of the panel on the front of the e/m apparatus. Finally, connect the negative terminal on the front panel of the apparatus to the negative terminal of the power supply. Check to make sure the multimeter is set to DC and set it on an appropriate scale to measure up to 2 A of current.

P5. Before proceeding with any measurements, make sure that you can identify the three separate circuits: one for the anode, one for the filament, and one for the magnetic field produced by the Helmholtz coils.

P6. Turn on the accelerating voltage (the knob above the black terminals) and set the voltage at about 250 V. Be sure to read the voltage from the multimeter (the one connected to the anode circuit, i.e., on the right side of the control panel), not the power supply.

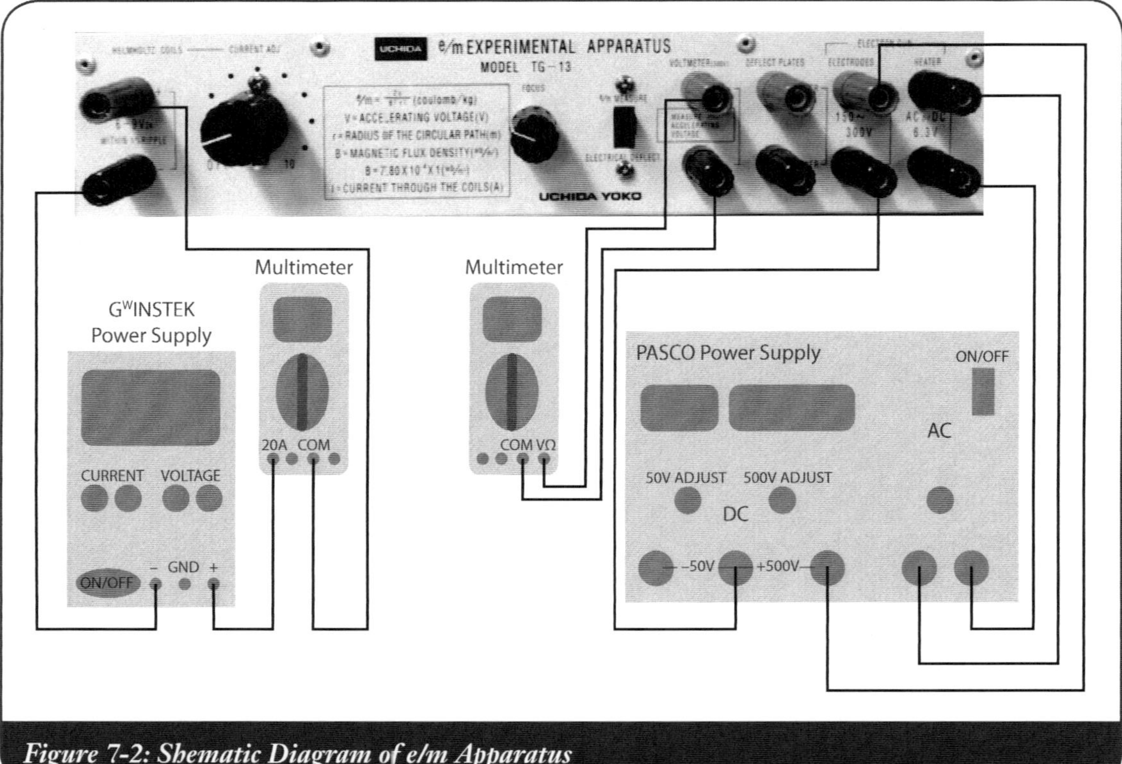

Figure 7-2: Shematic Diagram of e/m Apparatus

P7. Next turn the knob for the AC power supply (yellow terminals) to 6 volts (number 6 on the dial). This should result in a bluish beam of light that is easy to see. The filament may take a few minutes to heat up, so be patient. The beam should be parallel to the Helmholtz coils. If it isn't, turn the tube until it is. Do not remove the tube from its socket. The socket will turn with the tube.

ATTENTION!

Do not remove the tube from its socket.

P8. Now turn on the power supply that supplies current to the coils and adjust the current to somewhere between 0 A and 2 A. The electron beam will curve as the current increases. Adjust the current so that the radius of the circle described is somewhere between 0 and 5 cm.

P9. Carefully measure the radius of the circle described by the electron beam. Use the outside edge of the beam for your measurement. To avoid parallax errors, move your head to align the beam with its reflection on the mirrored scale behind the tube. Record the radius, R, of the beam on both the right and the left side. Also record the Helmholtz current, i, from the multimeter (the one connected in the series circuit that contains the terminals for the Helmholtz coils, on the left side of the control panel.)

P10. Repeat steps P8 and P9 for at least four different values of current in the Helmholtz coils. Each different current will result in a different radius for the circle described by the electron beam. As you adjust the current, keep the radius of the circle at 5 cm or less. (You should have a total of five different currents and the corresponding radii.)

P11. Repeat steps P8 through P10 using accelerating voltages of 275 V, 300 V, and 350 V.

P12. When you have completed all your measurements, turn off the power supply for the Helmholtz coils (the one connected to the left side of the control panel.) Next turn off the power to the filament (the dial that was set on 6). Then turn off the accelerating voltage (the high voltage that was set at around 250 V).

ATTENTION!

Make sure that the accelerating voltage is always set to at least 250 V whenever the power to the filament is on. Failure to do this could result in burning out the filament.

ATTENTION!

It's VERY important to turn off the power supply to the filament BEFORE turning off the power supply for the anode.

CALCULATIONS

C1. Calculate the average value of R for each setting in P9. Then use this value of R, along with the corresponding values of i and V to calculate e/m from Equation (5). (Remember that all quantities will need to be in SI units.)

C2. Calculate the average value of e/m from the five values obtained for each value of V. Then take the average of these four values of e/m (one for each voltage) to obtain a final experimental value of e/m.

C3. Compare your experimental value of e/m to the accepted value of 1.76×10^{11} coulomb/kg. Also calculate the uncertainty of your experimental value, and discuss whether or not your experimental value is consistent with the accepted value.

Notes

EXPERIMENT 8

The Current Balance and the Determination of the Permeability of Free Space

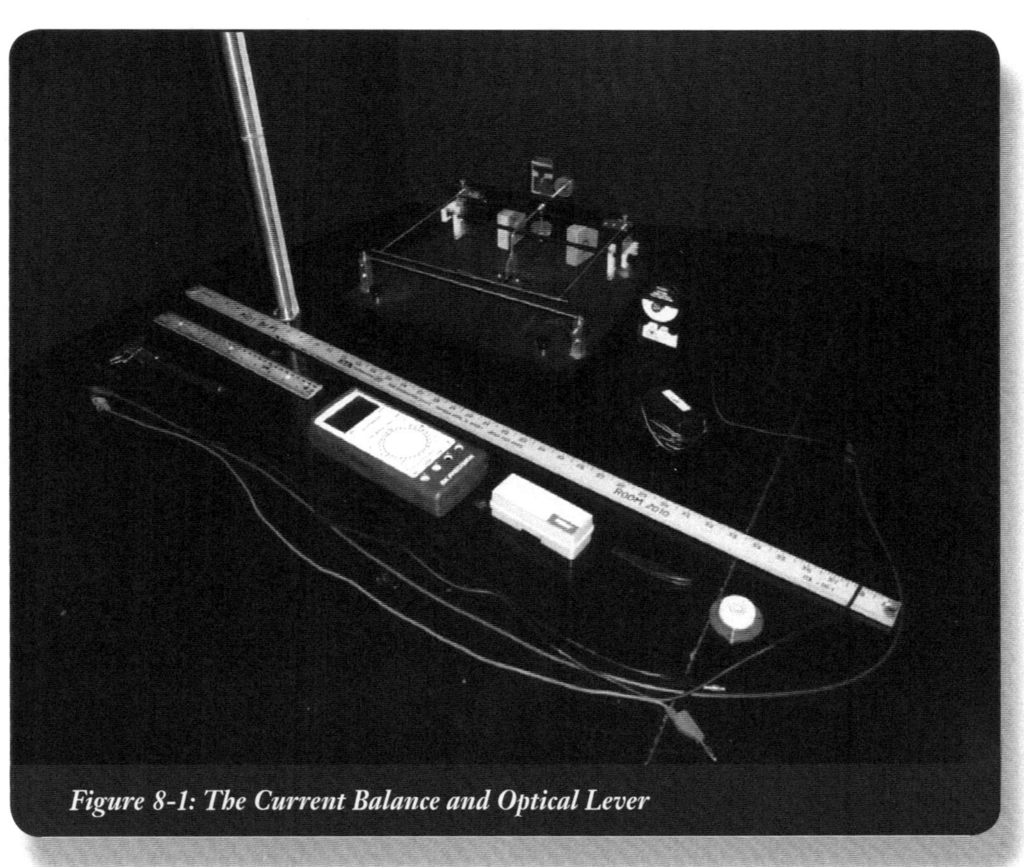

Figure 8-1: The Current Balance and Optical Lever

INTRODUCTION

Two parallel wires carrying equal currents i experience a mutual force, attractive if the currents are in the same direction, repulsive if in opposite directions. The magnitude of this force can be predicted by using Ampere's law to find the magnetic field **B** produced by one wire at the position of the other, then using the definition of **B** in terms of the force **F** produced on a current i. For wires of length L separated center to center by distance D, the result, in SI units,[11] is

$$F = \frac{\mu_0 i^2 L}{2 \pi R} \qquad (1)$$

CAUTION!

DO NOT TOUCH THE EXPERIMENTAL APPARATUS UNTIL YOU HAVE READ THE INSTRUCTIONS. The apparatus incorporates delicate knife edges which can be damaged easily. Please treat the apparatus with care. Do not look directly into the laser beam; to do so could cause severe eye damage. Look at beam reflections only. TAKE NO CHANCES.

where μ_0 is the permeability constant; it has the assigned value in SI units of exactly $4\pi \times 10^{-7}$ N/A^2. In this experiment a current balance is used to measure the force between two such wires, the measurement being aided by an optical lever to monitor small **displacements of one wire relative to the other upon application of** the current.

The entire apparatus is shown in Figure 8-1, and the current balance itself in Figure 8-2. The experiment begins with no current flowing through the balance; the counterweight behind the mirror is adjusted until the moveable wire comes to rest about 5 mm above the fixed wire when the frame is balanced on the knife edges. The counterweight under the mirror can be adjusted to make the period of oscillation of the moveable frame some convenient value, one or two seconds. The laser is then directed at the mirror on the frame, and the position of the reflected beam is noted on the vertical meter stick. Small weights are added to the moveable wire, driving the wire downward toward the fixed wire. Current is then sent through the wires, and since the current passes through the fixed and moveable wires in opposite directions, the magnetic force tends to counteract the gravitational force produced by the added weights. The current is adjusted until the reflected laser beam returns to the original position on the screen, indicating that the magnetic force between the currents has exactly compensated for the effect of the gravitational force. The process is repeated for different values of F and i, and a least-squares plot of F vs. i^2 is made. The slope of this plot gives the value of $\mu_0 L/2\pi R$ (Equation (1)). Knowing the values of L and R, μ_0 can be found and compared with the defined value.

PROCEDURE

P1. Carefully inspect the equipment and compare it to Figures 8-1 and 8-2. Notice that this is the apparatus used in the electrostatic balance experiment with metal rods now substituted for the metal plates. The same precautions listed there should be heeded to protect the equipment for this experiment. In addition, avoid bending the rods of the moveable frame and the stationary conductor. Begin by using the spirit level to level the base of the balance as you did in the electrostatic balance experiment.

P2. Before removing the wood blocks from beneath the movable frame, measure the length **L** of the portion of the frame which is to serve as the moveable wire (Figure 8-3); measure from

---
Note

[11] See, for example, Serway, *Physics for Scientists and Engineers* (3rd ed.), Section 30.2, or Giancoli, *Physics* (3rd ed.), Section 20–13.

center-to-center of its two supporting bars which extend from the knife edges. Also measure the distance **a** from knife edge to center of the front wire at each side; use the average of the two values in your calculations. Finally, use the vernier caliper to measure the diameter of the wires. (Here again it would be useful to measure each wire and take the average.)

P3. Following the precautions given in the electrostatic balance experiment, mount the frame onto the current balance. When properly mounted the vane attached to the movable frame will swing freely between the two damping magnets on the base of the balance. Using the beam-lift knobs, lift the frame off the knife edges and place the small wood blocks under the frame to hold it temporarily in place.

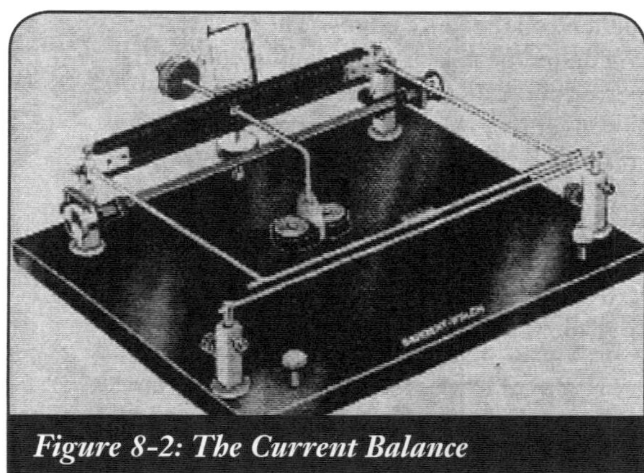

Figure 8-2: The Current Balance

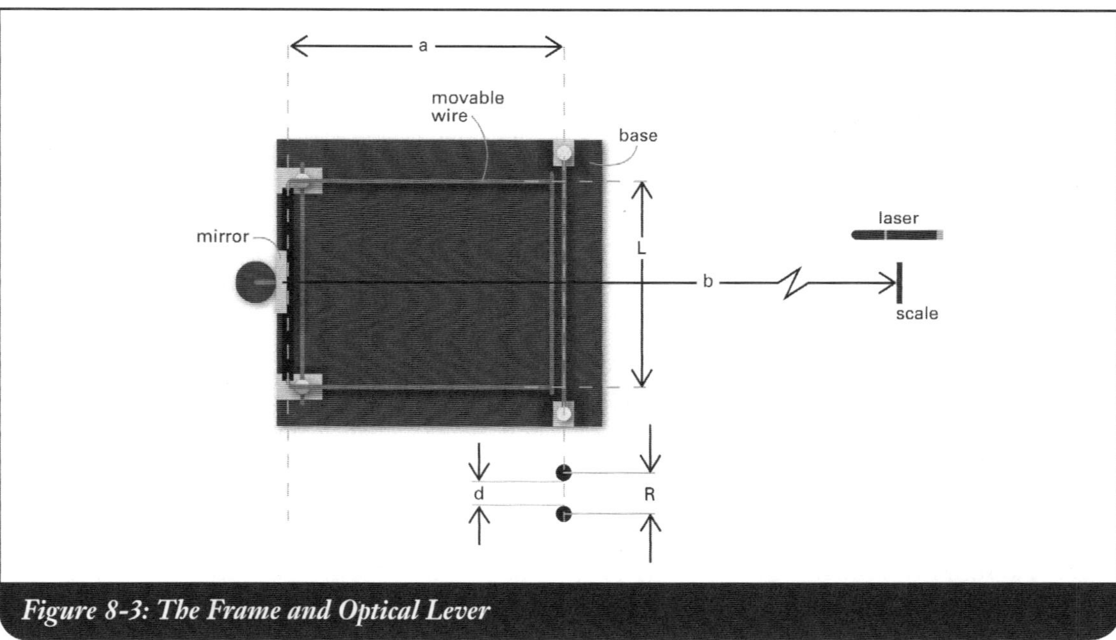

Figure 8-3: The Frame and Optical Lever

P4. Connect the Lab-Volt power supply, mounted on the apron of the laboratory bench (Figure 8-4), to the current balance, as shown schematically in Figure 8-5. The **RANGE** should be set to **"B,"** the rotating knob to **zero,** and the power switch set at **"off"** using the special key provided. The current balance is connected to the **yellow** terminals of the supply, i.e., A.C. current will be used. In order to minimize the effect of the fields produced by the connecting wires, connections to the binding posts on the balance should be made such that the connecting wires leave the posts at right angles to the conductors on the frame.

The use of alternating current rather than direct current eliminates the need to correct for the effect of the Earth's magnetic field on the current-carrying wires: for one direction of D.C. flow the Earth's field would tend to augment the magnetic force, but for the opposite direction would tend to reduce the net force. Using A.C. current avoids this difficulty. The effect of the Earth's field during one half of the A.C. cycle will be opposite to its effect during the other half, but because the current reverses direction 60 times each second the moveable frame cannot respond to the rapid changes, and an average position of balance is achieved.

Figure 8-4: The Lab-Volt Supply

P5. Remove the blocks from under the frame and gently lower it with the beam-lift knobs onto the knife edges. Adjust the counterbalance behind the mirror until the frame oscillates freely and comes to rest with the moveable wire about half a centimeter above the fixed wire; the counterweight below the mirror can be adjusted to make the period of oscillation some convenient value, on the order of one or two seconds. Check the alignment of the moveable and fixed wires by carefully placing a coin on the scale pan attached to the moveable wire; this will bring the moveable wire into contact with the fixed wire. Thumb screws on each front post permit either end of the fixed wire to be raised or lowered, and similar screws on the blocks at the rear of the balance permit the moveable wire to be moved forward or backward. Adjust the wires until they are parallel in all respects. Handle them gently and as little as possible, taking care not to bend them.

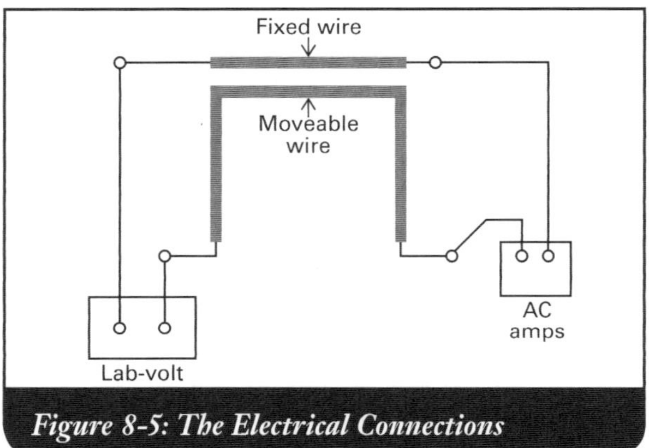

Figure 8-5: The Electrical Connections

P6. The pole with the meter stick attached to it will be used as an observing screen. Place it in the receptacle in the table and place the laser next to it.

P7. Remove the coin from the weight pan on the moveable wire and allow the moveable frame to come to equilibrium. Avoid excessive motion near the apparatus, as air currents can affect the results. Once equilibrium has been established, aim the laser at the mirror attached to the moveable frame. Adjust the angle of incidence of the laser beam until the reflected beam is seen on the vertical meter stick; if necessary the mirror can be tilted slightly by an adjusting screw located directly behind it. To check that the system is in its stable equilibrium position, gently lift the moveable frame off the knife edges using the beam-lift knobs, then again lower the frame back onto the knife edges. The laser beam should return to the same position on the screen; if not, check that the current balance is resting solidly on the laboratory table and that neither the laser nor the observing screen has been jarred from its initial position. Once you are certain stable equilibrium has been established, note the position of the laser beam on the vertical meter stick. step P8 will provide data to determine the separation between the moveable and fixed wires.

P8. Place the coin back on the scale pan, bringing the two wires into contact. Mark the new position of the laser beam on the observing screen. Record the distance **D** between the original position of the laser beam and the new position. Also measure **b,** the distance between the mirror and the observing screen. From simple geometry (Figure 8-6) it is seen that the distance between the top of the fixed wire and the bottom of the moveable wire, when at equilibrium, is given by

$$d = \frac{a\,D}{2\,b} \qquad\qquad (2)$$

Then the center to center distance **R** between the wires is the sum of **d,** as found from Equation (2), and the radii of both wires (Figure 8-3). The diameter of the wires measured in P2 is the same as the sum of the radii of the two wires so this value (**2r**) can be added to **d** to obtain **R.**

P9. Remove the coin from the weight pan and allow the moveable wire to return to its equilibrium position. Again avoid creating air currents near the balance. Using the forceps provided, add one of the small weights to the weight pan and wait for the frame to come to rest at a new equilibrium position. Slowly turn the adjustable knob clockwise on the Lab-Volt supply to provide current to the balance; the moveable wire will move back toward its original equilibrium position. Adjust the current until the laser beam returns to precisely the original position. Record the mass and the current.

P10. Repeat P9 for four more trials, adding an additional small weight each time. Record the total mass on the pan and the current required to bring the laser beam back to its original position for each trial. You should be able to use up to five weights total before the wires come into contact with one another. If this is not possible, you will need to readjust the apparatus so that the wires are farther apart and go back to P9 and restart your measurements. But be careful. If the spacing between the wires is too large, you will not be able to get enough current to bring the laser

ATTENTION!

Be careful NOT to move the apparatus until all measurements have been taken.

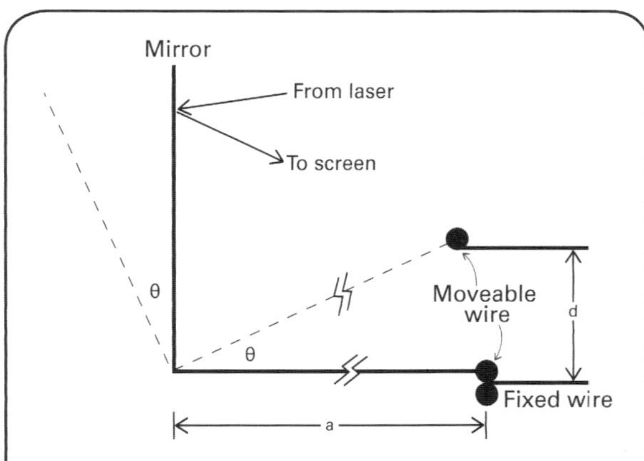

Figure 8-6: The Geometry of the Optical Lever (as the mirror and moveable wire rotate through an angle θ the laser beam reflected from the mirror rotates through 2θ; for small angles tanθ ≈ θ so θ ≈ d/a and 2θ ≈ D/b, where D is the linear distance through which the laser spot moves on the observing screen and b is the distance between the mirror and the screen (Figure 3); hence, d ≈ aD/2b)

beam back to its original position with five weights on the pan. Remember that the wire spacing must be kept constant as you change the amount of mass on the pan.

CALCULATIONS

C1. Using the data obtained in P9 and P10, make a least-squares plot of F vs. i^2; report the uncertainties in the slope and intercept as well as the values themselves. Include error bars on the data points in the graph, determined either through repeated measurement or by estimate based on your observations during the experiment.

C2. Calculate **R** from the value of **d** calculated in P8 and the value of **2r** obtained in P2 ($\mathbf{R} = \mathbf{d} + \mathbf{2r}$). From the slope of the line determined in C1 and the values of **L** (from P2) and **R,** calculate μ_0 using Equation (1) and compare your result with the defined value. Be certain to report the experimental uncertainty in your result.

C3. If you have previously performed Experiment 3, "The Electrostatic Balance and the Permittivity of Free Space," use the value obtained there for ε_0 and the result of C2 above to calculate $1/\sqrt{(\varepsilon_0 \mu_0)}$ and compare your answer with the expected result (what *is* the expected result?).

Notes

n = Index of refraction

$$n = \frac{c}{v} = \frac{\text{speed of light in vacuum}}{\text{speed of light in new medium}}$$

$n_{air} = 1.000$

Usually $n \geq 1.00$

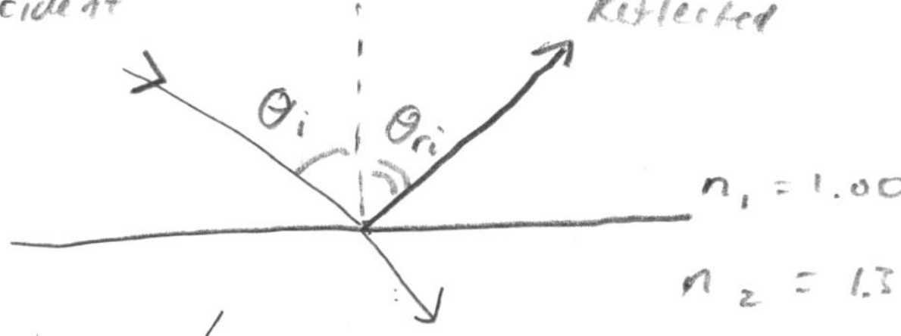

law of reflection $\theta_i = \theta_{ri}$

law of refraction = $n_1 \sin\theta_1 = n_2 \sin\theta_2$

- High to low, angle gets smaller
- Low to High, angle gets larger

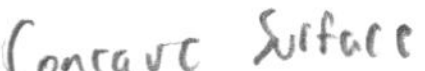

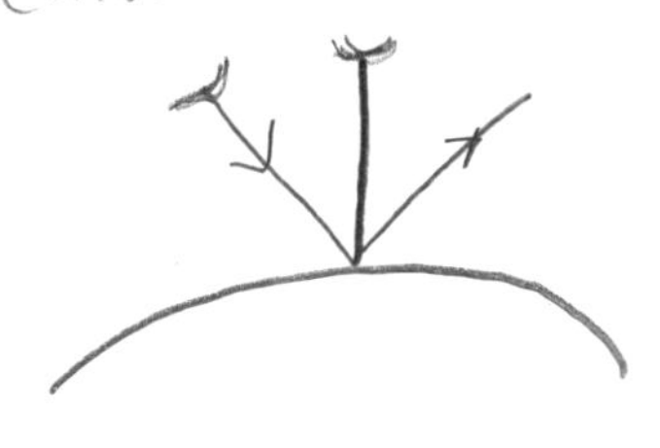

f = focal length

$R = 2f$

focal point

EXPERIMENT 9

Reflection and Refraction: The Basics

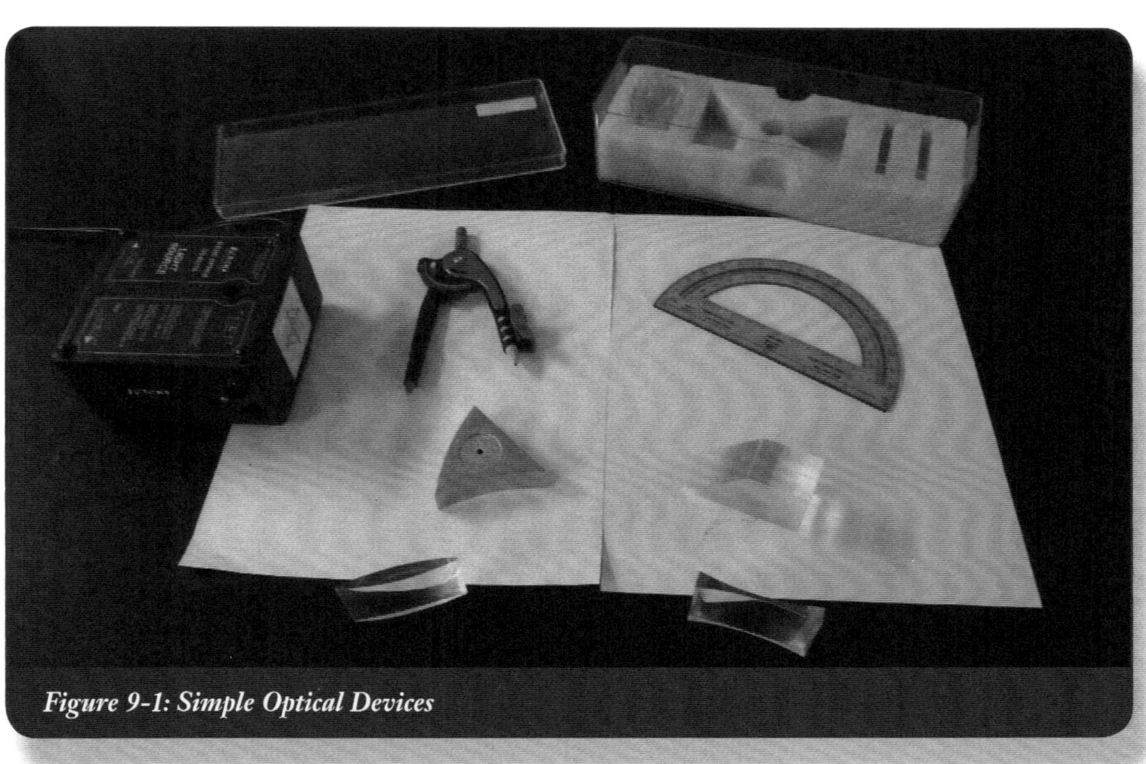

Figure 9-1: Simple Optical Devices

INTRODUCTION

When light waves encounter a boundary between two different media, some of the light is reflected, i.e., bounces off the surface and returns in the same media from which it came; and some of the light is refracted, i.e., continues into the new medium. The paths of the light reflected and refracted depend on the properties of the two media and on the shape of the boundary between the media. Here we investigate these properties for various types of boundary surfaces.

In order to discuss reflection and refraction in mathematical terms it is useful to think of light as a ray which is perpendicular to the wavefronts of the light and points in the direction of travel of the light. We can then discuss these phenomena in terms of the angles at which light approaches and then recedes from a boundary surface. It is customary to define these angles in terms of the <u>normal</u> to the boundary surface: a line drawn perpendicular to the boundary surface at the point of impact of the incident ray onto the surface. The angle of incidence (θ_i) is the angle between the incident ray and the normal, the angle of reflection (θ_{r1}) is the angle between the reflected ray and the normal and the angle of refraction (θ_{r2}) is the angle between the refracted ray and the normal. (See Figure 9-2 below.)

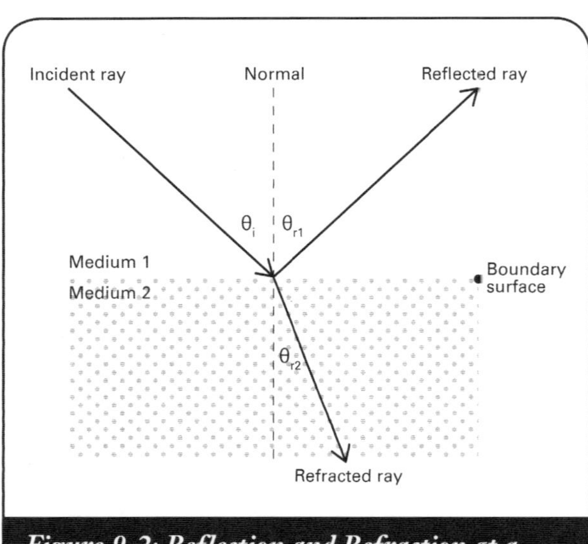

Figure 9-2: Reflection and Refraction at a Plane Boundary Surface

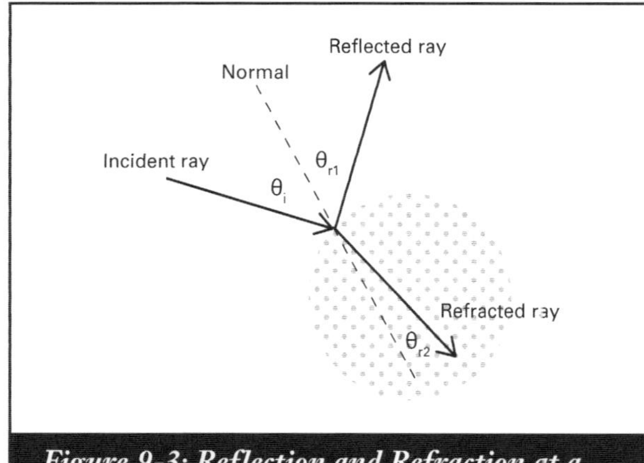

Figure 9-3: Reflection and Refraction at a Curved Boundary Surface

If the boundary surface between the two media is curved rather than straight as shown on the left, the same definitions can be applied but, in order to determine the normal line, it necessary to recall from geometry that the radius of a circle drawn from the center of the circle to any point P on the circle is perpendicular to the tangent to the circle at point P. A curved boundary surface can be thought of as a portion of a circle, so that the normal to the surface at any point of incidence is the radius of that circle that intersects the point of incidence. (See Figure 9-3.) According to theory the reflection of light is governed by the **law of reflection** which states that the angle of incidence is equal to the angle of reflection. Refraction is governed by the law of refraction, commonly referred to as **Snell's law:** $n_1 \sin \theta_1 = n_2 \sin \theta_2$, where n_1 and n_2 are the indicies of refraction of the first and second media respectively, θ_1 is the angle of incidence and θ_2 is the angle of refraction.

A light box, a triple reflecting surface, a prism and some simple lenses will be used to investigate these properties for various types of surfaces.

PROCEDURE

P1. Place the light box on a large piece of white paper with the slits which emit white light on the bottom. Adjust the slide on the front of the box so that one ray of white light is emitted from the box. Place the triangular mirror on the paper and aim the beam of white light at an angle to the flat side of the triangular mirror. Use a pencil to trace the incident ray and the reflected ray onto the paper. Also use the flat surface of the mirror as a guide to draw a line that represents the boundary surface. Use a protractor and a ruler to construct the normal to the surface at the point of impact of the light ray. (Remember that the normal line makes a 90° angle with the surface.) Repeat this process for two different angles of incidence. (That is, use a total of three different incident angles.)

P2. Now turn the triangular mirror so that the light ray hits the concave side. Trace the incident and reflected rays and draw a curved line along the mirror surface as before. Then remove the mirror and use the compass to construct the normal to the surface at the point of incidence. (It may be useful to extend this line out past the mirror surface.)

NOTE!

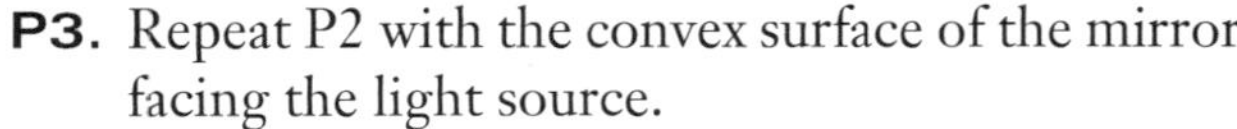

There are a couple of different methods that can be used to find the center of the circle and thus the normal. Use your knowledge of geometry or ask your instructor for a hint.

P3. Repeat P2 with the convex surface of the mirror facing the light source.

P4. Now adjust the slide on the light box so that five parallel rays of white light emerge from the box. Position the plane side of the triangular mirror so that the light rays strike it at normal incidence. (See Figure 9-4 (a).) What do you notice about the reflected rays? Now turn the triangular mirror so that the light rays hit its concave surface and so that the center ray hits the center of the mirror at normal incidence. (See Figure 9-4 (b).)

Trace the incident and reflected rays and the surface of the mirror. Also use the compass as before to locate the center of curvature of the mirror and mark it on the paper. The point at which the rays reflected from the mirror intersect is called the focal point of the mirror. Mark this point also.

P5. Now turn the mirror so that the light rays hit the convex side. (See Figure 9-4 (c).) The reflected rays now spread out as they emerge from the mirror. In fact they appear to be coming from a common point behind the mirror. Trace the incident and reflected rays and

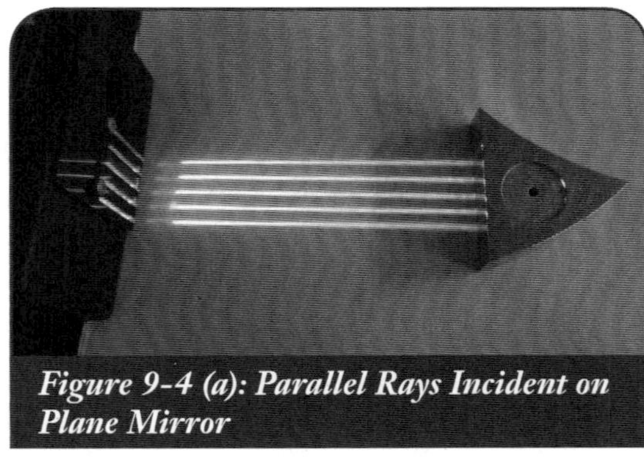

Figure 9-4 (a): Parallel Rays Incident on Plane Mirror

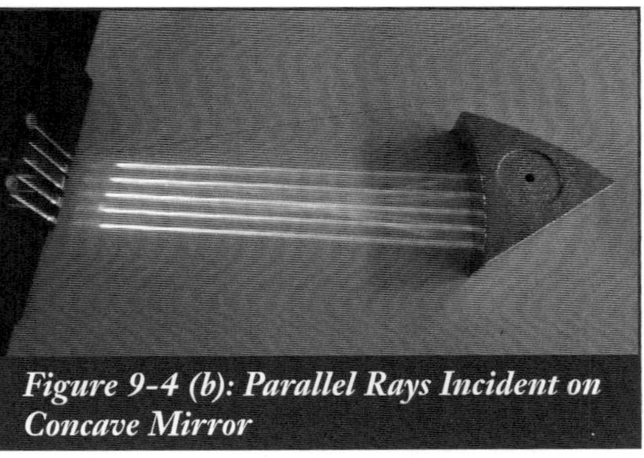

Figure 9-4 (b): Parallel Rays Incident on Concave Mirror

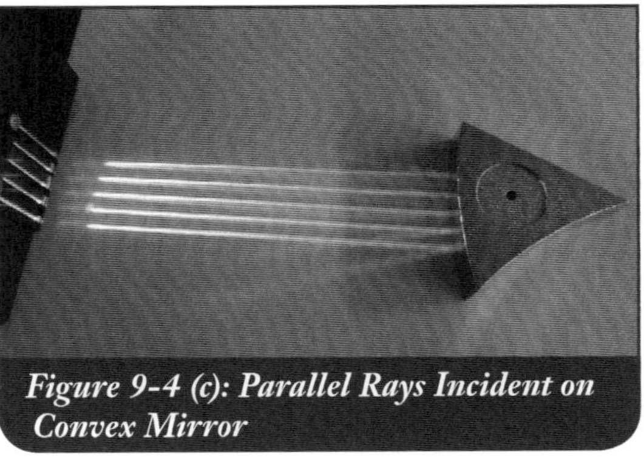

Figure 9-4 (c): Parallel Rays Incident on Convex Mirror

the surface of the mirror. Then remove the mirror and use a ruler to extend the reflected rays backward so that you can find the point at which they intersect behind the mirror. Mark this point. It can be thought of as the virtual focal point. As before use the compass to locate the center of curvature of the mirror and mark it on the paper.

P6. Readjust the light box so that again only one ray of white light is emitted and place the light box and the prism on the white paper so that the ray of white light hits one of the flat sides of the prism at an angle of incidence of at least 45°. (See Figure 9-5.) Instead of concentrating on the reflected ray, this time we will examine the refracted ray as the light travels first from the air into the prism and then from the prism back out into the air. Trace the incident ray and mark the spot where it hits the prism. Also trace the ray that emerges from the prism and mark the exit point for this ray. Remove the prism and draw a line from the point of entrance to the point of exit for the prism. (This is the refracted ray inside the prism.) Use a ruler and a protractor to construct the normal to <u>each</u> surface of the prism at the entrance and exit points. (Remember that the normal for each surface must make a 90° angle with the surface.)

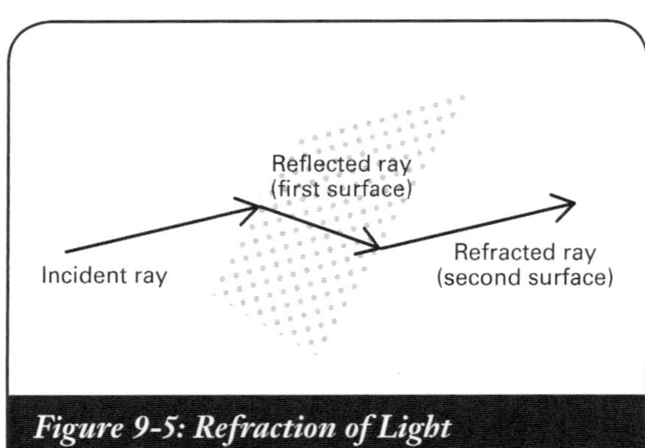

Figure 9-5: Refraction of Light

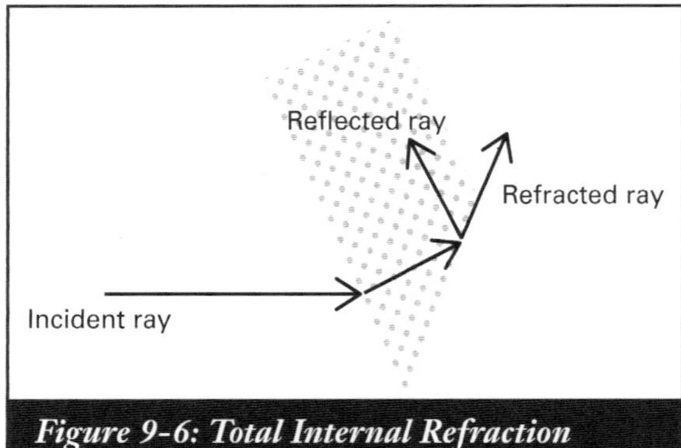

Figure 9-6: Total Internal Refraction

P7. Now position the light source and the prism as shown in Figure 9-6. (Be careful not to aim the incident light too close to the pointed tip of the prism.) Examine the incident ray, the reflected ray and the refracted ray for the <u>second</u> surface, i.e., the one <u>outside</u> the prism. Now slowly rotate the prism in a clockwise direction until this refracted ray disappears. At this point **total internal reflection** occurs and the reflected ray will be at a 90° angle to the incident ray. Carefully mark the point of entrance of the light to the <u>first surface</u> of the prism. Also mark the point of exit of the

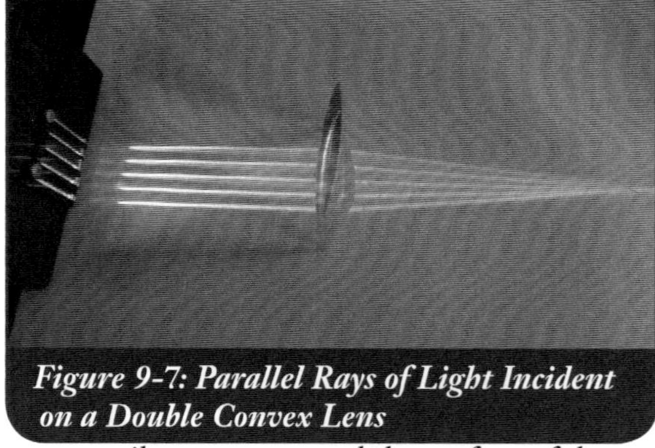

Figure 9-7: Parallel Rays of Light Incident on a Double Convex Lens

light from the <u>second surface</u> of the prism. Use your pencil to trace around the surface of the prism. Then remove the prism and draw a line connecting the entrance and exit points of the light. This is the incident ray for the second surface. Use a ruler and protractor to construct the normal for this surface and measure the angle of incidence. This is called the **critical angle.**

P8. Now readjust the light box to again allow five parallel rays of white light to emerge. Place the light box on the white paper and place the hand-held double convex lens in the path of the light rays. Position the lens at a right angle to the rays of light. (See Figure 9-7.) What do you notice about the light rays after they pass through the lens? The point where the rays converge is called the focal point of the lens. Trace the incident and refracted rays and mark the **focal point** for this lens. Measure the focal length of the lens (the distance from the center of the lens to the focal point.)

P9. Replace the double convex lens with the double concave lens (Figure 9-8) and again notice what happens to the light rays after passing through the lens. Trace the rays and extend the refracted rays backward to the virtual focal point behind the lens. Mark this point and again measure the focal length of the lens. Note that the focal length of a diverging lens such as this one is labeled as <u>negative</u> since the focal point is a virtual one and occurs behind the lens.

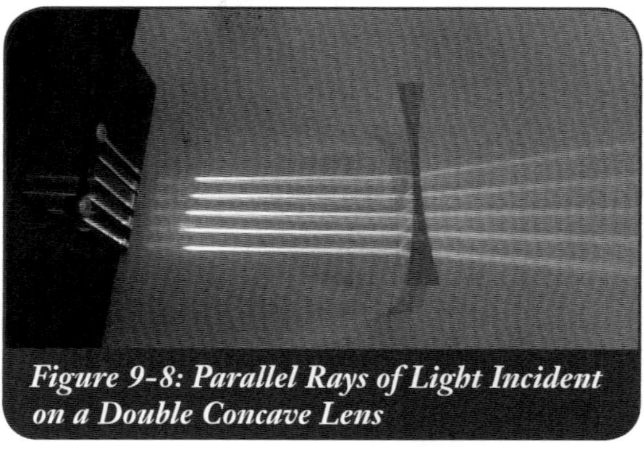

Figure 9-8: Parallel Rays of Light Incident on a Double Concave Lens

CALCULATIONS

C1. Measure the angle of incidence and the angle of reflection for light hitting the flat surface of the triangular mirror for each of the three different angles of incidence in P1. Calculate per cent difference between the two. Do your results agree, within experimental error with the law of reflection as stated in your text?

C2. Repeat C1 for the concave surface of the triangular mirror used in P2.

C3. Repeat C1 for the convex surface of the triangular mirror used in P3.

C4. Measure the distance from the center of the concave mirror in P4 to the focal point. This is the **focal length (f).** Also measure the radius of the circle that fits the curve of the mirror. This is called the **radius of curvature (R)** of the mirror. Do you notice a relationship between **f** and **R**? What is the focal length for the flat surface of the mirror? What is the radius of curvature for that surface?

C5. Measure the distance from the center of the convex mirror in P5 to the virtual focal point behind the mirror. This can be thought of as the focal length of the convex mirror. (It will be a <u>negative</u> number since the focal point is a virtual one.) Also measure the radius of curvature for this surface and compare it to the focal length.

C6. Measure the angle of incidence and the angle of refraction for each surface of the prism in P6. Note that the refracted ray from the first surface becomes the incident ray for the second surface. What do you notice about the bending of light at each surface? Does it bend toward or away from the normal in each case? What about the direction of the light emerging from the second surface as compared to the direction of the original incident ray? Use Snell's law to determine the index of refraction of the prism for each refraction. Compare these values to each other and to the theoretical value of 1.5.

C7. Use Snell's law to calculate the expected critical angle for light reflected from the back surface of the prism in P7. (Assume the index of refraction for the prism is 1.5.) Compare this to the experimentally measured value for the critical angle.

C8. Do you notice any relationship between the focal lengths of the double convex and the double concave lens as measured in P8 and P9? Does this make sense to you based on your observations of these two lenses?

Spherical Mirror
↳ Primarily reflect light
- Concave mirror

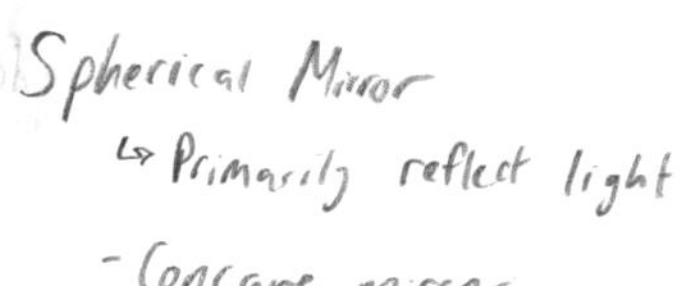

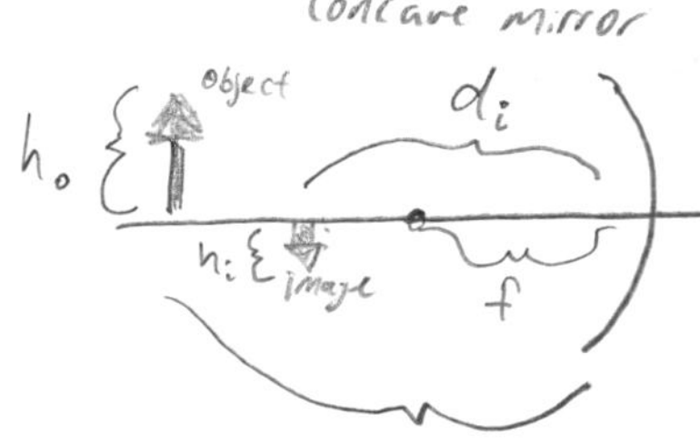

d_0 = distance to object

Concave → f is +
Convex → f is −

Lenses let light pass through (refract)

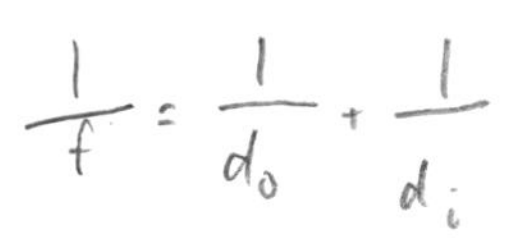

f = focal length
d_0 = distance to object
d_i = distance to image

$$\frac{1}{f} = \frac{1}{d_0} + \frac{1}{d_i}$$ } Mirror equation ←→ Same for lenses

Magnification
$$M = \frac{h_i}{h_o}$$

★ if image is upside down, h_i is negative

$$M_{theoretical} = \frac{-d_i}{d_0}$$

★ d_i is positive if reflection image is in front of the mirror

Reflection and Refraction:
Curved Mirrors, Thin Lenses, and Image Formation

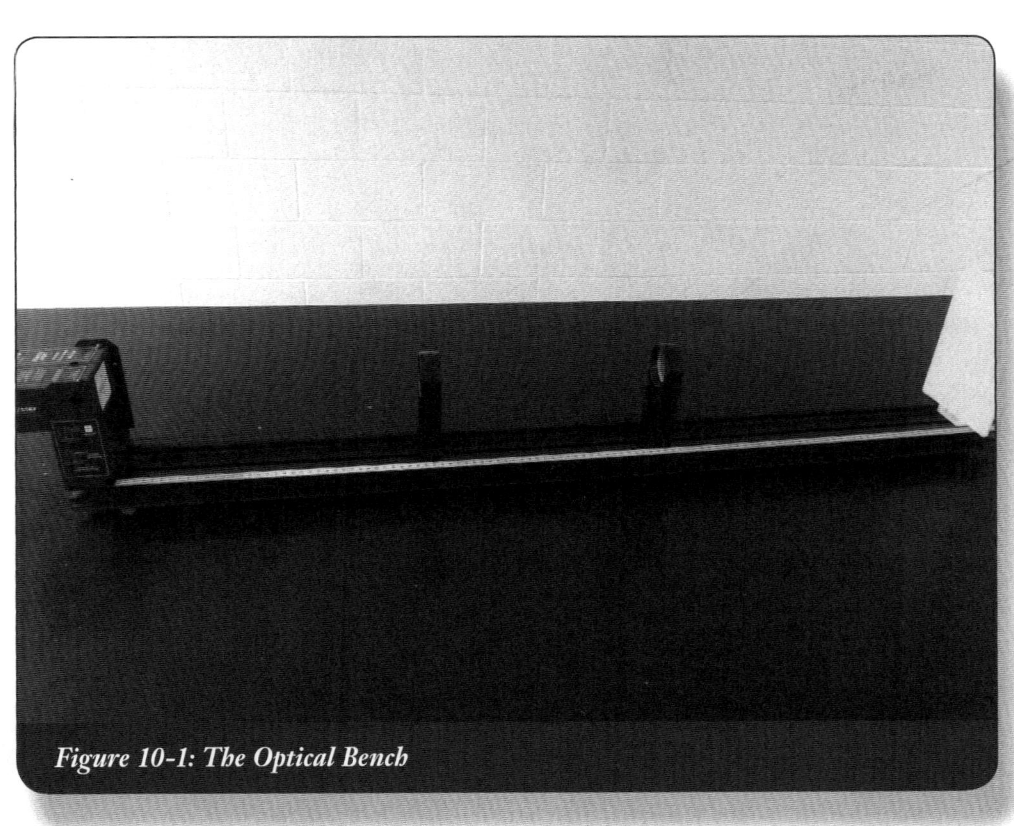

Figure 10-1: The Optical Bench

INTRODUCTION

The principles of geometrical optics which were investigated in the previous experiment can be used to obtain equations which determine the focal lengths and the size and position of the images for various types of lenses and mirrors.

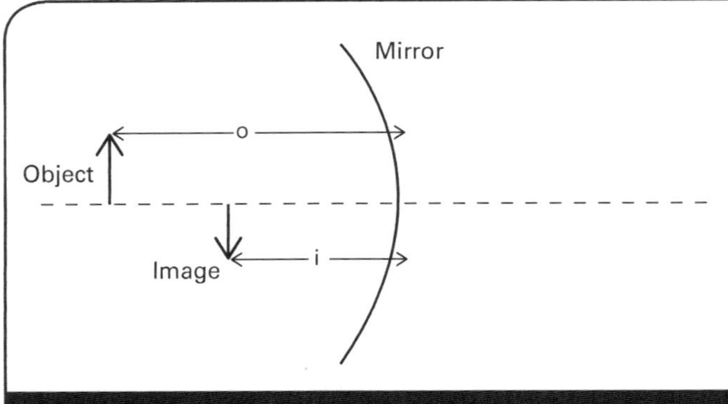

Figure 10-2: Image Formation by a Curved Mirror

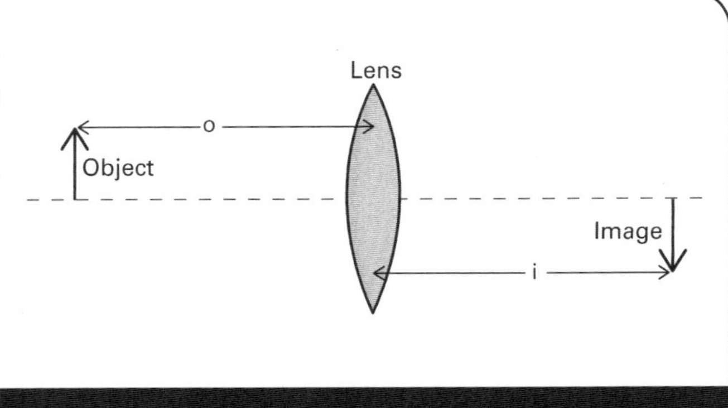

Figure 10-3: Image Formation by a Lens

The light box is used as an object and is placed in front of an optical element (a mirror or lens) on the light bench. A screen is then placed on the bench in the appropriate position to obtain an image. (See Figures 10-1 and 10-2.) The distance between the object and the optical element is labeled **o** and the distance between the optical element and the image is labeled **i**. The focal length, **f,** of the mirror or lens can be shown to be given by the equation:

$$\frac{1}{f} = \frac{1}{o} + \frac{1}{i} \tag{1}$$

This is called the <u>lens equation</u>, but is actually valid for both thin lenses and mirrors. In fact, this equation can be used to locate an image or to find the focal length for any type of thin lens or mirror (or combination) and for any configuration, if the appropriate values of **o, i,** and **f** are used.

For the simple cases illustrated in Figures 10-2 and 10-3 where the object is placed in front of the lens or mirror and a real image is formed on a screen, **o, i,** and **f** are all positive. However, in some cases an image cannot be focused on a screen but is seen by the eye, which "imagines" where the light appears to be coming from. This is called a virtual image. If an image appears where actual light rays exist after passing through an optical element the image is said to be real and the image distance is taken as positive. If the image appears in a region where there are no actual reflected or refracted light rays, the image is virtual and the image distance is taken as negative. This means that for mirrors, the image distance will be taken as positive when the image is on the same side of the lens as the object and negative when it is on the opposite side. For lenses the opposite is true: the image distance is positive for images on the opposite side of the lens from the object and negative on the same side of the lens as the object.

For converging mirrors or lenses, the focal length is positive and for diverging mirrors or lenses the focal length is taken as negative. The object distance is taken to be positive except in the case where the image from one lens is used as a virtual object for a second lens as in the last part of this experiment where the focal length of a lens combination will be determined.

One common use of lenses and mirrors is to produce an increase or reduction in the size of an image as compared to an object. This is defined quantitatively as magnification (**M**). The magnification of an optical element can be defined as the ratio of the image size to the object size. That is, if we define the object size as **h** and the image size as **h'**, then

$$M = \frac{h'}{h} \qquad (2)$$

This equation can be used along with direct measurements of **h** and **h'** to determine an experimental value of the magnification. If the image is upright, **h'** is taken to be positive and if the image is inverted, **h'** is taken as negative, so that a negative value of **M** corresponds to an inverted image. Notice that if the image is larger than the object the absolute value of **M** will be greater than one and if the image is smaller than the object it will be less than one.

Application of the principles of reflection and refraction shows that if the appropriate signs are included as described above, the magnification is also related to the object and image distances through the equation

$$M = -\frac{i}{o} \qquad (3)$$

If two lenses are combined the effect of the combination can be determined by taking the image for the first lens as the object for the second lens and using two successive applications of the lens equations. If the image from the first lens happens to be behind the first lens (i.e., on the opposite side to what would normally be expected), then it is said to be a virtual object for the second lens and the object distance for that lens is appropriately taken to be negative.

PROCEDURE

P1. Place the concave mirror on the optical bench in front of the light source so that the distance between the two is between 30 cm and 60 cm and place the half-screen between the mirror and the light source. (See Figure 10-4.) Move the screen back and forth until you see a well focused image on the screen. Record the positions of the light source (the object), the mirror and the screen (the image). Also measure the size of the object and the image and record these as h and h' respectively.

Figure 10-4: Image Formation by a Concave Mirror

Now turn the mirror around so that the convex side faces the light source and again translate the screen back and forth along the bench. Can you find a position of the screen which allows a sharp image to be formed?

Remove the screen and stand behind the light source and look at the mirror. You will see an image that is the same size as the object and appears to be behind the mirror. This is produced by reflection from the plane glass surface that is in front of this side of the mirror. You will also see a smaller spot of light just below the first image. Move the mirror back and forth until this spot of light becomes a clear image of the source. This is called a virtual image. This image cannot be focused on a screen because it is not actually formed by light rays but rather it is the point where the light rays appear to come from.

P2. Now remove the mirror and replace it with the +100 mm focal length converging lens and replace the half-screen with the full screen. Leave the light source turned off and allow light from a distant source (such as the lamp from one of the other lab groups) to enter the lens from the side opposite the light source. (See Figure 10-5.) Move the lens back and forth until a clear image of the source can be seen on the screen. Record the positions of the lens and screen.

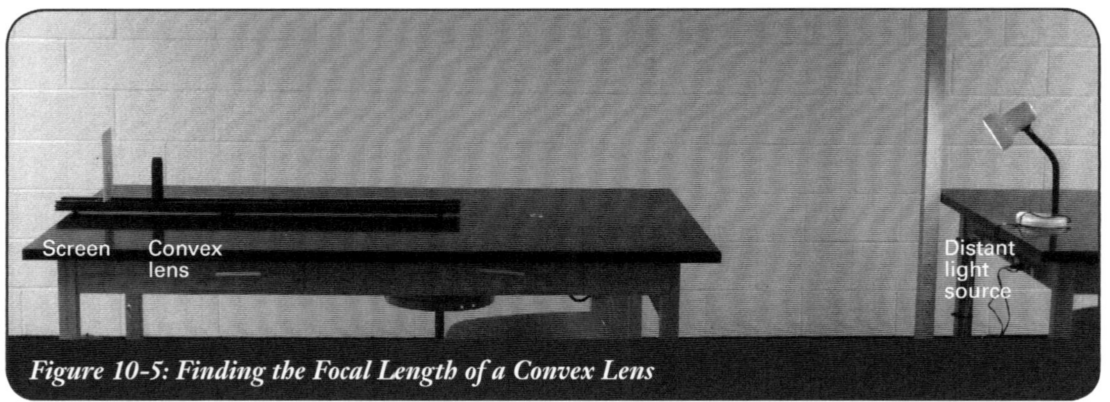

Figure 10-5: Finding the Focal Length of a Convex Lens

P3. Now turn on the light source and place the screen on the opposite side of the lens from the source. (See Figure 10-6.) The object (light source) and the screen should be about 80 cm apart. Record the positions of the object and the screen. Move the lens back and forth until you find two positions where a clear image is formed on the screen. (When the lens is close to the source a large image will be visible; when the lens is closer to the screen the image will be much smaller.)

Record each of these positions. Also measure and record the size of the object and the image.

Next move the screen a few centimeters closer to the object. Record this position and again find and record two positions of the lens which produce a clear image on the screen. Repeat for three more different positions of the screen in relation to the object, so that you have a total of 10 different positions of the lens (2 for each of 5 different positions of the screen).

Replace the lens marked +100 mm with the one marked –150 mm. Move the lens back and forth to see if you can obtain an image on the screen. Next position the lens at about the 50 cm mark and stand <u>behind</u> the light source and look back toward the lens. You will see two images: one image is small and seems fixed, and the other is larger and moves as you move your eyes close to the back of the source. The first (fixed) image is virtual and the second is real. The second image comes from a reflection off the first concave surface of the lens. Try to project it onto a piece of paper.

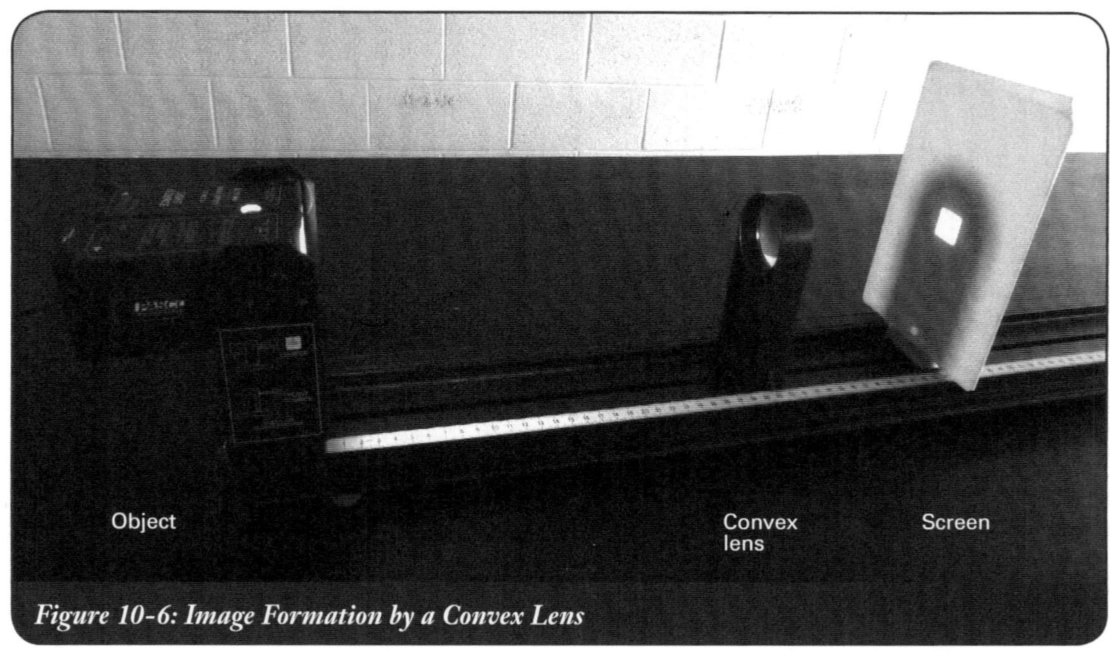

Figure 10-6: Image Formation by a Convex Lens

P4. Remove the –150 mm lens and again place the +100 mm lens between the object and the screen. Move the lens back and forth until a clear image is formed on the screen. Record the positions of the object, lens and screen. This image will become the **virtual object** for the diverging lens which will be added next. Also record the size of the object and the image.

Now place the diverging lens just in front of the converging lens and move the screen back and forth until a clear image is again formed on the screen. (See Figure 10-7.) Record this new position of the screen. Also record the size of the new image.

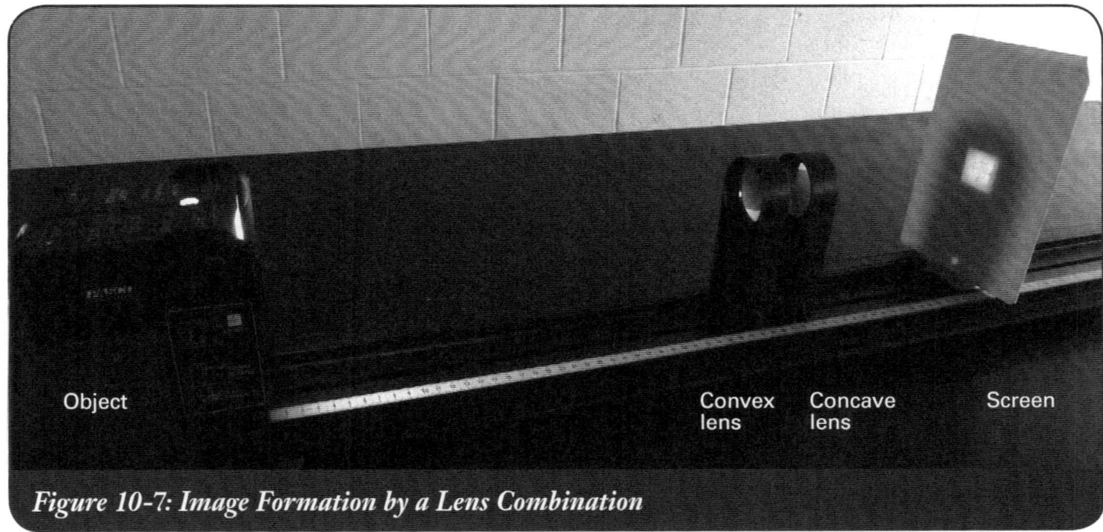

Figure 10-7: Image Formation by a Lens Combination

CALCULATIONS

C1. Use the positions of the light source, screen and mirror that you recorded when forming an image with the concave mirror in P1 to calculate the object distance and the image distance. Then calculate the focal length using $1/f = 1/o + 1/i$.

Compare this experimental value for f to the focal length as marked on the mirror. Calculate the theoretical value of the magnification for the concave mirror using $M = -i/o$ and compare it to the experimental value obtained from the measured values of h and h' ($M = h'/h$). Be sure to use the appropriate signs for the distances and the heights.

C2. Calculate the difference in the positions of the lens and the screen for the distant object focused by the +100 mm convex lens in P2. This value is the focal length of the lens. (Why?) How does it compare with the marked value?

C3. Make a table of the object and image distance for each of the configurations of object, lens and screen that produced a clear image using the +100 mm convex lens in P3. Plot a graph of $1/o$ vs. $1/i$. Find the x and y intercepts of this graph. Each of these will give a value of $1/f$. (Examine the lens equation to see why.)

Calculate f from each of these and compare these values to each other and to the value marked on the lens.

Calculate and compare the experimental and theoretical values of magnification for <u>just one</u> of the positions of the screen (two different positions of the lens).

C4. Use the distance between the first image in P4 (the one formed by the convex lens before adding the concave lens) and the diverging lens as the <u>negative</u> object distance for the diverging lens, and use the distance between the final image and the diverging lens as the image distance to calculate the focal length of the diverging lens from the lens equation.

QUESTIONS

1. For each type of mirror or lens, under what conditions is a real image formed and under what conditions is the image virtual. Are these results consistent with what you saw in the previous experiment. Explain.

2. For each type of mirror or lens, under what conditions is the image larger than the object and under what conditions is it smaller?

3. For which situations is the image upright and for which situations is it inverted? Is there a relationship between whether the image is real or virtual and whether it is upright or inverted? Explain.

Notes

R = dist between slit & screen

a = size of slit

λ = wave length of light

θ = angle between center & point of interest

y = dist. along screen from center

m = integer

Red 700 nM

Green 500 nM

EXPERIMENT 

Interference and Diffraction

Figure 11-1: Laser, Slit Accessory, and Screen

INTRODUCTION

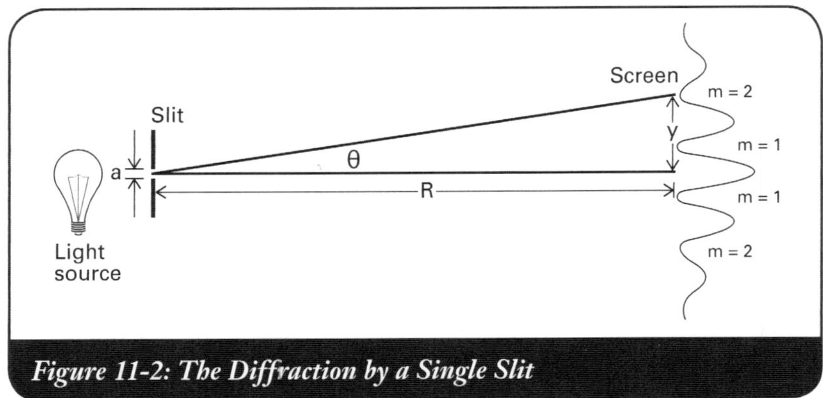

Figure 11-2: The Diffraction by a Single Slit

When light is incident on a slit or obstacle whose dimensions are of a magnitude similar to the wavelength of the incident light, the light spreads out from the slit (or obstacle) or diffracts, and different parts of the wave are combined according to the principle of superposition to form a diffraction pattern: a series of bright and dark regions corresponding to constructive and destructive interference respectively. If a screen is placed at some distance R away from the slit or obstacle as in Figure 11-1, the positions of the minima are given by the equation:[12]

$$\sin \theta = \frac{m\lambda}{a} \qquad (1)$$

where λ is the wavelength of the light, a is the width of the slit, θ is the angle between a line drawn from the slit to the center of the pattern and one drawn from the slit to the particular minimum being examined, and m is the order (that is the count for the particular minimum). The first dark region on either side of the central peak is the first order minimum ($m = 1$), the second dark region on either side corresponds to $m = 2$, etc. From Figure 11-2 we see that if y is the distance between the central peak and the mth minimum, then θ for that order is given by

$$\tan \theta = \frac{y}{R} \qquad (2)$$

(R is the horizontal distance from the center of the slit to the center of the diffraction pattern on the screen.) For small orders θ is generally sufficiently small that we can use the approximation $\tan \theta \approx \sin \theta \approx \theta$, so that combination of Equations (1) and (2) gives

$$\frac{y}{R} = \frac{m\lambda}{a} \qquad (3)$$

If two slits instead of one are used, the pattern is similar. For the double slit pattern the positions of the **maxima** shown in Figure 11-3 are given by[13]

$$\sin \theta = \frac{n\lambda}{d} \qquad (4)$$

_______ *Notes*

[12, 13] Giancoli, *Physics, Principles with Applications* (5th ed.) p. 734, or Wolfson, Pasachoff, *Physics with Modern Physics for Scientists and Engineers* (3rd ed.), p. 991.

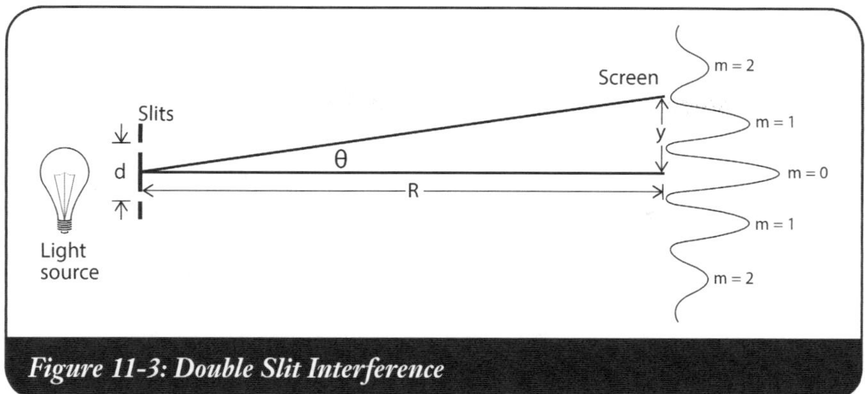

Figure 11-3: Double Slit Interference

where λ is the wavelength as before, **d** is the distance between the two slits and **n** is the order of the interference maximum; i.e., **n** = 0 for the central maximum, **n** = 1 for the first maximum on either side of the central one, etc. θ is now the angle between the line from the center of the slits to the **n**th maximum and the line connecting the center of the slits to the center of the interference pattern. Examination of Figure 11-3 shows that Equation (2) holds here as well.

If light is incident upon a large number of equally spaced slits as with a diffraction grating, the resulting pattern is similar to the double slit pattern but with the bright lines being narower and sharper. The pattern can be shown[14] to obey Equation (1), with **a** being the space between slits, λ the wavelength and **m** the order of the interference **maximum.**

PROCEDURE

P1. Place the laser on the light bench and connect the power supply to it. Place the single slit accessory in front of the laser and attach a sheet of white paper to the screen with tape and place it in front of the single slit accessory. Plug in the power supply and turn on the laser. Rotate the disc on the slit accessory so that the variable slit is directly in front of the laser and adjust the laser so that a clear diffraction pattern is formed. Start with the laser aimed at the narrow end of the variable slit and gradually rotate the disc so that the slit becomes increasingly wider.

WARNING!

Do not look directly into the laser beam. To do so could severely damage your eye.

P2. Now position the disc so that the slit marked **a** = 0.04 mm is directly in front of the laser. Adjust the laser if necessary to obtain a clear diffraction pattern on the screen. Use the meter stick to measure and record the distance, **R,** from the slit to the screen. Make sure your measurement is made to the slit and not to the center of the slit's holder since these positions are a few millimeters apart. Use the vernier caliper to measure and record the distance between the two dark lines nearest the center of the diffraction pattern. These are the 1st order minima. Also measure and record the distance between the second order minima on either side.

Note

[14] Giancoli, *Physics, Principles with Applications* (5th ed.) p. 736, or Wolfson, Pasachoff, *Physics with Modern Physics for Scientists and Engineers* (3rd ed.), p. 978.

P3. Now rotate the disk so that the slit with **a** = 0.08 mm is in front of the laser. Again measure and record the distance between the minima on either side of the central maximum for **m** = 1 and **m** = 2. It should not be necessary to remeasure **R** unless you needed to move the slit or the laser in order to increase the size of the pattern.

P4. Remove the holder with the single slit accessory and replace it with the holder on which the double slit accessory is mounted. Rotate the disc so that the variable double slit is positioned in front of the laser and adjust the laser so that a clear interference pattern is visible on the screen. Start at the end where the slits are closer together and slowly rotate the disc counterclockwise so that the distance between the slits gradually increases.

P5. Readjust the disc so that the pair of double slits labeled **a** = 0.08 mm, **d** = 0.25 mm is positioned in front of the laser. Use the vernier to measure the distance between the left and right maxima for as many different values of **n** as you can. Record these distances. Also measure and record **R** (the distance between the slits and the screen) if it is not the same as in the previous steps.

P6. Rotate the disc again so that the double slits labeled **a** = 0.08 mm, **d** = 0.50 mm is positioned in front of the laser. Again measure the distance between maxima for each value of **n** as in P5.

P7. Now rotate the disc and examine the interference patterns for the two sets of slits marked **a** = 0.04 mm (**d** = 0.25 mm and **d** = 0.50 mm respectively as in P5 and P6).

P8. Readjust the disc so that the laser is aimed at the first of the sets of multiple slits, i.e., the one marked 2. Examine the pattern formed by this set of slits and then turn the disk to the multiple slits marked 3, 4, and 5 successively. What happens to the pattern as the number of slits is increased?

P9. Finally remove the holder with the slit accessory and position the diffraction grating so that the laser shines through it and produces a diffraction pattern on the screen. Measure and record the distance between maxima on the left and right side of the central maximum for as many values of m as you can. Measure and record the distance between the diffraction grating and the screen.

P10. Remove the diffraction grating and reinsert the holder with the single slit accessory in front of the laser. Adjust the disc to examine the diffraction patterns produced by the circular apertures and the two dimensional patterns (squares, hexes, dots, and holes).

P11. Remove the red laser and replace it with the green laser. Observe the diffraction patterns obtained with a single slit, a double slit and the diffraction grating. How does the spacing of the lines in these patterns compare to the patterns from the corresponding elements with the red laser? Does this result agree with what you would expect from Equation (4)?

CALCULATIONS

C1. How does the diffraction pattern change as the width of the single slit is gradually increased in P1? Is this consistent with what you should expect according to Equation (1)? Explain.

C2. Using the distance between minima obtained for the single slit in P2, divide this distance by 2 to obtain the value of **y** in Equation (3). Using the known value of 0.04 mm for **a**, calculate the wavelength of the laser light from Equation (3). Do this calculation for both **m** = 1 and **m** = 2. Calculate the percent difference for the two values of λ obtained. Also calculate the average value of λ. This average value will be used for the wavelength in the remainder of the experiment.

C3. Use the measurements made in P3 to obtain **y** for each order for the diffraction pattern from the 0.08 mm slit. How does this pattern compare to the previous one? Use Equation (3) along with the values obtained for **y** and the wavelength obtained from C2 to determine the width of this slit. You will obtain two values—one for each value of **m.** Compare these experimentally obtained values for **a** to the marked value of **a** = 0.08 mm.

C4. How do the results for diffraction for a line compare to those for a slit of the same size?

C5. Describe how the double slit interference pattern changes as the distance between the slits gradually increases. Are your observations consistent with Equation (4)? Explain.

C6. Use the measurements made in P5 to determine the value of θ for each value of **n** for the double slit marked **a** = 0.08 mm, **d** = 0.25 mm. Use these values and the experimental value for the wavelength obtained earlier along with Equation (4) to obtain an experimental value for **d** (the spacing between the slits) for each value of **n.** Take the average of these values for **d** and compare this average to the **d** = 0.25 mm value marked on the slits.

C7. How does the interference pattern obtained in P6 compare to the pattern for the more closely spaced slits of P5? Can you explain this relationship in terms of Equation (4)?

C8. How do the patterns for the smaller slit width seen in P7 compare to those for the larger slit width in P5 and P6? Can you explain this in terms of the equations for interference and diffraction previously discussed?

C9. Use the measurements obtained in P9 and the experimental value obtained earlier for the wavelength of the laser light along with Equation (1) to calculate the value, **a**, of the spacing between slits in the diffraction grating used in P10. (Note that the angles involved here are probably too large for the small angle approximation to be valid.) Compare this value of **a** to the one marked on the grating.

Notes

EXPERIMENT 12

Optical Spectra

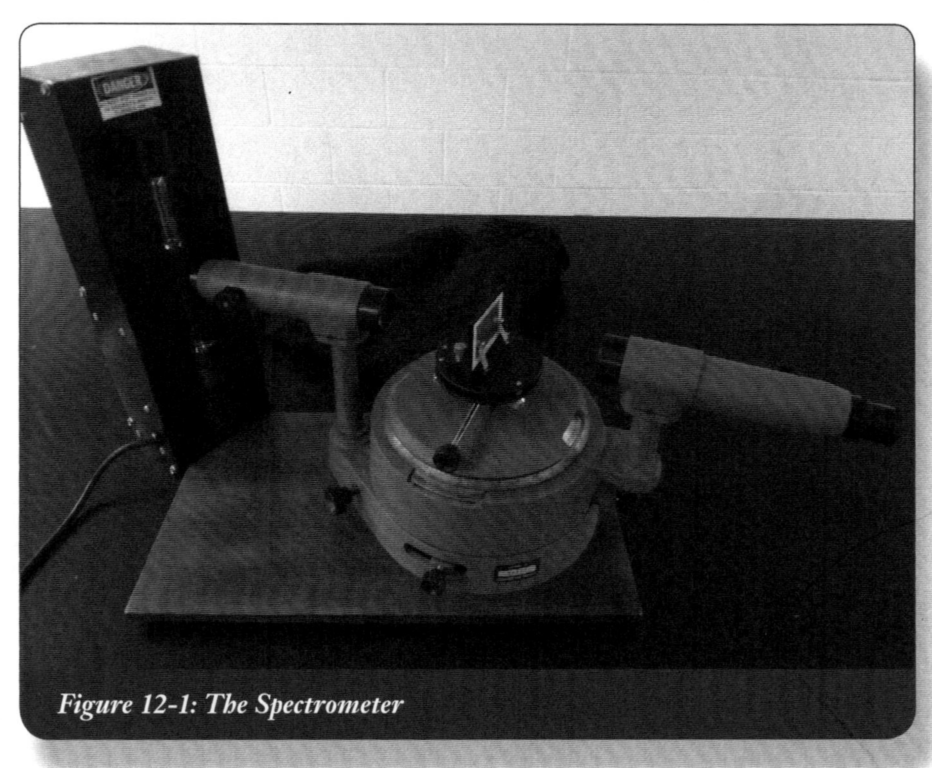

Figure 12-1: The Spectrometer

INTRODUCTION

In Experiment 11 we examined the pattern of bright lines formed by the interference of laser light incident on a diffraction grating. Recall that the bright lines occurred at positions which correspond to constructive interference and those positions are determined by the equation

$$m \lambda = d \sin \theta \qquad (1)$$

where **d** is the spacing between slits in the diffraction grating, λ is the wavelength of the incident light and **m** is the order of the diffraction maximum (**m** = 1, 2, 3, ...). The laser light source used in Experiment 11 was highly monochromatic; that is, it contained only one specific wavelength. If the light incident on a diffraction grating contains more than one wavelength, there will be a diffraction pattern similar to the one obtained with the laser for each wavelength of light present. Notice that the angle in Equation (1) depends on the wavelength, so the bright lines of constructive interference will appear in slightly different positions for different wavelengths of incident light. So a light source which contains several different wavelengths of light will produce a diffraction pattern consisting of a series of bright lines of various colors. The light sources used here will be gas discharge tubes containing two different gas elements: mercury and helium. The set of wavelengths emitted by such a tube is characteristic of the particular element used, so the pattern of lines in the diffraction pattern, i.e., the **spectrum** can be used to identify the element. Each chemical element has its own characteristic spectrum consisting of a particular set of wavelengths.

If the spectrum for a particular source is known, Equation (1) can be used, along with measurements of the positions of the spectral lines produced by the source, to determine the slit spacing of a diffraction grating. Conversely, if the slit spacing for a diffraction grating is known the wavelengths contained in the spectrum of a given source can be calculated from Equation (1).

Here we will use the mercury discharge tube as a known source from which we will determine the slit spacing of a diffraction grating. We will then use the helium source as an unknown source and calculate the expected wavelengths for this source based on the previously calculated value of the slit spacing.

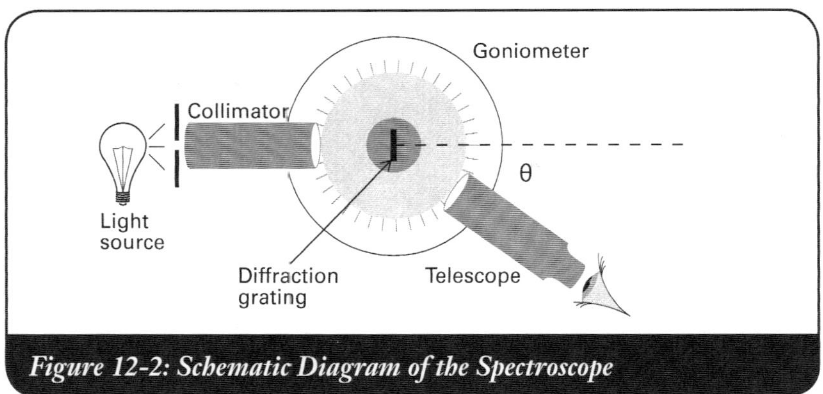

Figure 12-2: Schematic Diagram of the Spectroscope

The apparatus used here is the **spectroscope,** which consists of a narrow slit at one end of a collimator, a table onto which the diffraction grating is placed, a telescope which is free to rotate to locate the spectral lines, and a goniometer, which is a circular vernier scale used to measure the angle for each spectral line. The gas discharge tube is placed directly in front of the slit and the telescope is rotated until the cross hairs line up with the spectral line being measured. (See Figures 12-1 and 12-2.) We will use only the first order maxima here so Equation (1) becomes

$$\lambda = d \sin \theta \qquad (2)$$

PROCEDURE

P1. Place the mercury source in front of the slit and position the diffraction grating on the table so that light from the source hits it at normal incidence. Position the telescope directly opposite the source and adjust the slit so that a clear, narrow blue line is aligned with the crosshairs.

P2. Next rotate the telescope to the right until you see a series of bright colored lines that match the sequence in Table 1. This is the first order spectrum. If you rotate the telescope further to the right you should see the same series of lines repeated in the second order spectrum, and even further to the right you may be able to see the third order spectrum. Also rotate the telescope to the left of the center and locate the first and second and possibly the third order spectrum on that side.

P3. Once you have familiarized yourself with the spectrum and have located and identified the lines of the first order spectrum, you can begin taking measurements. Move the telescope so that the crosshairs are positioned over each of the spectral lines in the first order spectrum to the right of the center and for each one record the angle measurement from the vernier scale. (See the appendix for a description of how to read the vernier scale.) Then do the same for each of the lines to the left of the center. Make a table of the left angle and the right angle for each of the lines. (Choose a scale on one side or the other of the spectroscope and use it for the measurements of <u>all</u> the spectral lines on both sides of the central maximum. Do <u>not</u> use both scales.)

P4. Now replace the mercury source with the helium source and repeat P2 and P3, making a correspondence with as many lines as you can from Table 2 for the helium spectrum. Some of the lines in the table will not be visible. Use only the ones that are, but make sure you establish a correspondence with the sequence of lines you see and the ones in the table before making any measurements. You can use the wavelengths and the relative intensities to help you determine which ones to use.

Table 1: Spectral Mercury[15]

Color	λ (nm)
purple	435.833
green	546.074
yellow	576.960
yellow	579.066

Table 2: Spectral Lines for Helium[16]

Color	λ (nm)	Intensity
purple 390–455 nm	388.8650	500
	396.4729	20
	402.6191	50
	412.0820	12
	438.7929	10
	447.1479	200
	447.1680	25
blue 455–492 nm	471.3146	30
	492.1931	20
green 492–577 nm	501.5678	100
	504.7740	10
yellow 577–597 nm	587.5620	500
	587.5970	100
red 622–780 nm	667.8150	100
	706.5190	200
	706.5710	30
	728.1350	50

Notes

[15] *Handbook of Chemistry and Physics*, 74th ed., 1993–1994, Sec. 10, p. 57.

[16] Ibid, Sec. 10, p. 39.

CALCULATIONS

C1. Find the difference between the angles recorded on the right and on the left of the center for each of the spectral lines of mercury obtained in P3. This is twice the angle θ in Equation (2). Divide by 2 to obtain θ.

C2. Use Equation (2) along with the wavelengths from Table 1 and the values of θ from C1 to calculate a value of **d** (the slit spacing) for each value of θ. Take the average of these to obtain the **d** value that will be used in the next part to calculate the wavelengths in the spectrum of helium.

C3. Calculate 2θ and then θ for each of the spectral lines of helium as you did for the lines of mercury in C1.

C4. Use the values of θ from C3 and the average value of **d** from C2 to in Equation (2) to calculate the experimental value of the wavelength for each of the spectral lines of helium. Compare these to the values shown in Table 2.

NOTE!

If the scale goes through 360° and starts over between the two measurements you will have to account for this in your calculation. Also you will need to convert the angles obtained in degrees, minutes and seconds to degrees and fractions of degrees before using them in Equation (2). For example if the reading is 172°22'30" as in one of the examples in the appendix, first convert 30 seconds to 30/60 minutes or 0.5 minutes, so that the reading becomes 172°22.5'. Then convert 22.5 minutes to 22.5/60 degrees or 0.375 degrees. So the final reading is 172.375°.

APPENDIX

There are two different types of vernier scales so each will be discussed separately here:

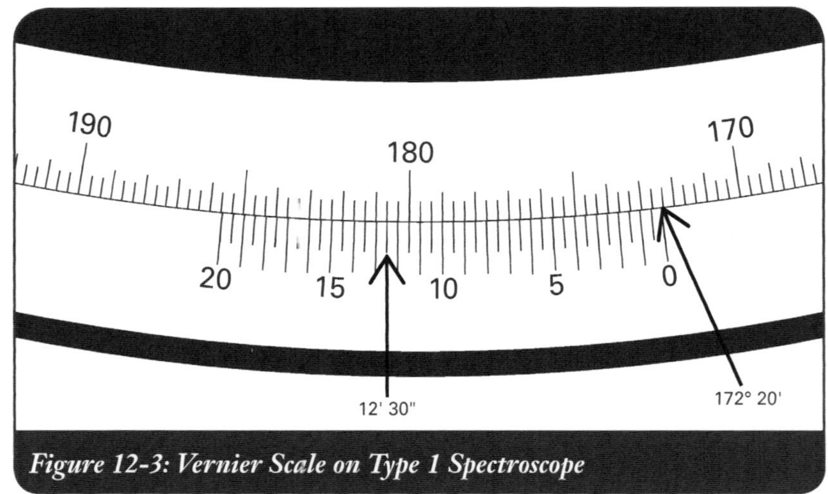

Figure 12-3: Vernier Scale on Type 1 Spectroscope

TYPE 1

First look at the top scale. You will notice that the largest marks occur at five degree intervals with numerical labeling at each ten degree mark and an unlabeled five degree mark in between each of these. The space between adjacent five degree marks is then divided by smaller marks into five equal segments, so these next smaller marks correspond to single degrees. Now notice that the space between two adjacent degree marks is divided into three sections. Therefore each of these corresponds to ⅓ degree or, since one degree = 60 minutes, these smallest marks correspond to ⅓ (60 minutes) = 20 minutes. So the first mark past a whole degree represents 20 minutes and the second represents 40 minutes.

The first step in reading the vernier scale is to locate the zero on the bottom scale between two of the smallest marks on the upper scale. The smaller of these gives the first approximation of the scale reading. For example, in Figure 12-3 this reading would be 172°20'.

Next scan to the left until you find a mark on the bottom scale that exactly matches up with a mark above it on the top scale. Notice that the bottom scale goes from 0 to 20 minutes with the largest marks corresponding to 5 minutes, the next smaller marks to individual minutes and the smallest marks corresponding to one-half minute or 30 seconds. (There are sixty seconds in a minute.) Take the reading on this bottom scale for the mark that lines up with one on the top scale. In the example in Figure 12-3 this would be 12'30'.

Finally take the reading from the first scale and add it to the reading for the second scale to obtain the final reading. For the example in Figure 12-3 this would give 172°20' + 12'30' = 172°32'30".

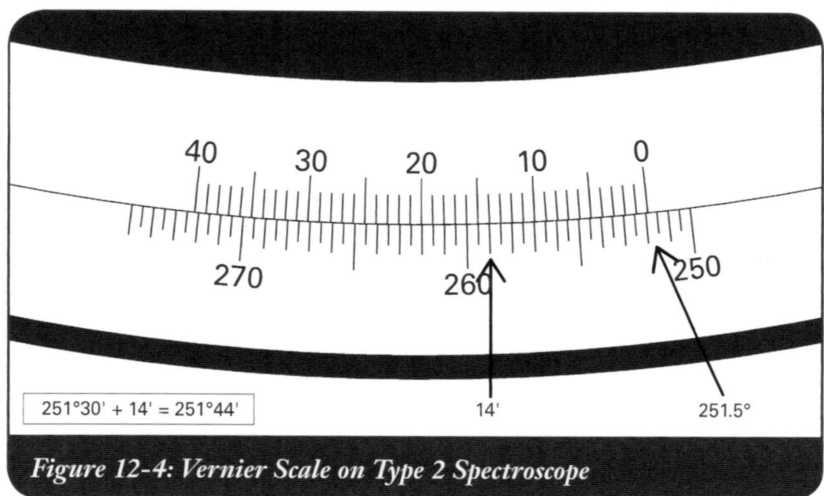

Figure 12-4: Vernier Scale on Type 2 Spectroscope

TYPE 2

First look at the bottom scale. You will notice that the largest marks occur at five degree intervals with numerical labeling at each ten degree mark and an unlabeled five degree mark in between each of these. The space between adjacent five degree marks is then divided by smaller marks into five equal segments, so these next smaller marks correspond to single degrees. Those marks are further subdivided into half-degree intervals by the smallest marks.

The first step in reading the vernier scale is to locate the zero on the top scale between two of the smallest marks on the bottom scale. The smaller of these gives the first approximation of the scale reading. For example, in Figure 12-4 this reading would be 251.5° or 251°30'. (Recall that one degree equals 60 minutes so one-half degree equals 30 minutes.)

Next scan to the left until you find a mark on the top scale that exactly matches up with a mark below it on the bottom scale. Notice that the top scale goes from 0 to 30 minutes with the largest marks corresponding to 5 minutes and the smaller marks corresponding to individual minutes. Take the reading on this top scale for the mark that lines up with one on the bottom scale. In the example in Figure 12-3 this would be 14'.

Finally take the reading from the first scale and add it to the reading for the second scale to obtain the final reading. For the example in Figure 12-3 this would give 251°30' + 14' = 251°44'.

Notes

Notes

Notes

Notes

Notes